Plant Resource Allocation

This is a volume in the

PHYSIOLOGICAL ECOLOGY series
Edited by Harold A. Mooney

A complete list of books in this series appears at the end of the volume.

Plant Resource Allocation

Edited by

Fakhri A. Bazzaz
Department of Organismic and Evolutionary Biology
Harvard University
The Biological Laboratories
Cambridge, Massachusetts

John Grace
Institute of Ecology & Resource Management
School of Forestry & Ecological Sciences
The University of Edinburgh
Edinburgh EH9 3JU, United Kingdom

Academic Press

San Diego   London   Boston   New York   Sydney   Tokyo   Toronto

Cover photo credit: Rich Knavel (Sequoia tree).

This book is printed on acid-free paper. ∞

Academic Press
a division of Harcourt Brace & Company
525 B Street, Suite 1900, San Diego, California 92101-4495, USA
http://www.apnet.com

Academic Press Limited
24-28 Oval Road, London NW1 7DX, UK
http://www.hbuk.co.uk/ap/

Library of Congress Cataloging-in-Publication Data

Plant resource allocation / edited by Fakhri A. Bazzaz, John Grace.
      p.   cm. -- (Physiological ecology series)
  Includes bibliographical references and index.
  ISBN 0-12-083490-1
  1. Plant resource allocation. I. Bazzaz, F. A. (Fakhri A.)
II. Grace, J. (John), date. III. Series: Physiological ecology.
QK717.P58 1997
581.7--dc21   97-12027
                                   CIP

PRINTED IN THE UNITED STATES OF AMERICA
97 98 99 00 01 02 EB 9 8 7 6 5 4 3 2 1

Contents

1. Allocation of Resources in Plants: State of the Science and Critical Questions
Fakhri A. Bazzaz

2. The Fate of Acquired Carbon in Plants: Chemical Composition and Construction Costs
Hendrik Poorter and Rafael Villar

3. Resource Allocation in Variable Environments: Comparing Insects and Plants
Carol L. Boggs

4. Biomass Allocation and Water Use under Arid Conditions
Hermann Heilmeier, Markus Erhard, and E.-Detlef Schulze

5. Organ Preformation, Development, and Resource Allocation in Perennials
Monica A. Geber, Maxine A. Watson, and Hans de Kroon

6. Optimality Approaches to Resource Allocation in Woody Tissues
R. M. Sibly and J. F. V. Vincent

11. Allocation Theory and Chemical Defense
Manuel Lerdau and Jonathon Gershenzon

12. Toward Models of Resource Allocation by Plants
John Grace

Contributors

Numbers in parentheses indicate the pages on which the authors' contributions begin.

David Ackerly (231), Department of Biological Sciences, Stanford University, Stanford, California 94305

Fakhri A. Bazzaz (1), Department of Organismic and Evolutionary Biology, Harvard University, Biological Laboratories, Cambridge, Massachusetts 02138

Carol L. Boggs (73), Center for Conservation Biology, Department of Biological Sciences, Stanford University, Stanford, California 94305; and Rocky Mountain Biological Laboratory, Crested Butte, Colorado 81224

T. J. de Jong (211), Institute of Evolutionary and Ecological Sciences, University of Leiden, 2300 RA Leiden, The Netherlands

Hans de Kroon (113), Department of Terrestrial Ecology and Nature Conservation, Agricultural University, 6708 PD Wageningen, The Netherlands

Markus Erhard (93), Department of Plant Ecology, University Bayreuth, D-95440 Bayreuth, Germany

Monica A. Geber (113), Section of Ecology and Systematics, Cornell University, Ithaca, New York 14853

Jonathon Gershenzon (265), Max Planck Institute for Chemical Ecology, Washington State University, Institute of Biological Chemistry, Pullman, Washington 99164

John Grace (279), Institute of Ecology & Resource Management, School of Forestry & Ecological Sciences, The University of Edinburgh, Edinburgh EH9 3JU, United Kingdom

Hermann Heilmeier (93), Department of Plant Ecology, University Bayreuth, D-95440 Bayreuth, Germany

Michael J. Hutchings (161), School of Biological Sciences, University of Sussex, Brighton, Sussex BN1 9QG, United Kingdom

Peter G. L. Klinkhamer (211), Institute of Evolutionary and Ecological Sciences, University of Leiden, 2300 RA Leiden, The Netherlands

Manuel Lerdau (265), Department of Ecology and Evolution, State University of New York, Stony Brook, New York 11794

Hendrik Poorter (39), Department of Plant Ecology & Evolutionary Biology, Utrecht University, 3508 TB Utrecht, The Netherlands

Edward G. Reekie (191), Biology Department, Acadia University, Wolfville, Nova Scotia, Canada BOP IXO

E.-Detlef Schulze (93), Department of Plant Ecology, University Bayreuth, D-95440 Bayreuth, Germany

R. M. Sibly (143), School of Animal and Microbial Sciences, The University of Reading, Whitenights, Reading RG6 2AJ, United Kingdom

Rafael Villar (39), Department Biología Vegetal y Ecología, Universidad de Córdoba, 14004 Córdoba, Spain

J. F. V. Vincent (143), School of Animal and Microbial Sciences, The University of Reading, Whitenights, Reading RG6 2AJ, United Kingdom

Maxine A. Watson (113), Department of Biology, Indiana University, Bloomington, Indiana 47405

Preface

A fundamental and vital activity of plants is the uptake, processing, and allocation of resources from the environment. These resources include basic materials such as CO_2, water, and nutrients as well as manufactured materials such as sugars, proteins, and defensive chemicals. The allocation of resources to various plant activities (growth, reproduction, and defense) must be under strong selection. For plants that have repeatedly encountered specific environmental circumstances, the process of allocation and trade-offs between various activities and functions also must have been honed by natural selection. Allocation patterns and allocation theory play a pivotal role in life history evolution and in functional plant ecology. In recent years, the use of economic theories and models and mathematical formulations has contributed to our understanding of allocation in organisms. It is assumed that allocation takes place in such a way as to maximize fecundity and overall lifetime fitness.

Because of the central role that allocation plays in evolutionary and functional ecology, we brought together leading contributors to the study of resource allocation. These exceptional chapters are certain to generate much discussion and a great deal of excitement.

The book is organized into 12 chapters, prepared by very energetic authors, modified from their presentations to include, discuss, and consider the comments received from the audience. Each chapter was then externally reviewed and read by both editors. The suggestions made by various readers were also incorporated in the current text. The editors are extremely happy with the final product. We believe it treats this critical subject in an up-to-date way. We feel that the chapters in this book will generate much enthusiasm and much discussion among students of ecology and evolution.

We thank all of the authors for their participation and excellent chapters and Dr. C. Crumly and his staff at Academic Press for seeing the book through publication.

FAKHRI A. BAZZAZ
JOHN GRACE

1

Allocation of Resources in Plants: State of the Science and Critical Questions

Fakhri A. Bazzaz

I. The Plant Functions as a Balanced System: Flexibility of Allocation

Plant biologists have long recognized that in order for a plant to complete its life cycle, it must function as a balanced system in terms of resource uptake and use (e.g., Mooney, 1972; Agren and Ingestad, 1987). Communication between carbon gaining and nutrient and water gaining parts of the plant is assumed to be rapid and efficient. Resources obtained from the environment and manufactured in the plant are allocated to various plant parts and functions (growth, reproduction and defense) in accordance with this view. Many models and experiments make the fundamental assumptions that natural selection has molded plant allocation in economics terms (see Field, 1991). Investment in any function should be terminated when the "return" on that investment falls below the investment (Mooney and Gulmon, 1979; Bloom *et al.*, 1985). For example, investment in the enzyme Rubisco is reduced when light is limiting, and investment of nitrogen shifts toward chlorophyll to obtain the more limiting resource, light. Carbohydrate status seems to coordinate the balance between photosynthesis and respiration (Amthor, 1995).

Shifts in allocation patterns under changing environmental conditions have been experimentally proven to maximize plant growth (e.g., Robinson and Rorison, 1988; Mooney *et al.*, 1988; Hirose, 1987). At the physiological level, it is assumed that plants allocate resources so that pool sizes within the plant remain constant (Schulze and Chapin, 1987). It is also assumed that environmental limitations, or excesses, that reduce resource use will also reduce resource uptake (Chapin, 1991a,b). However, there is some evidence that the uptake and transport of some nutrients can exceed demands for growth (Schulze, 1991). Is this merely luxury uptake to deprive neighbors? How quick is the adjustment to achieve this balance? How much of these resources remain in the active pool? What is the cost of their storage, if any? The speed of adjustment to the prevailing environmental conditions varies among plants and is a critical aspect of their strategies. It follows from this expectation that plants with relatively high nonstructural materials (e.g., herbaceous annuals) are more flexible in redeployment than those with relatively high structural materials (e.g., long-lived trees) and that plants which occupy habitats with highly variable environments have a higher flexibility of allocation and redeployment. They must track their environment, i.e., quickly change their resource allocation in response to environmental change. However, there is only limited experimental evidence for this situation at this time. *Plantago major*, which is more common in repeatedly disturbed habitats, is more responsive, in terms of allocation, to nutrient pulses than its congener *P. rugelii*, which is common in less disturbed habitats (Miao *et al.*, 1991). In *P. major* there is a significant increase in allocation to reproduction, and there is a negative correlation between vegetative and reproductive biomass. With a nutrient pulse, *P. major* increased its leaf relative growth rate (RGR_l) and decreased its root relative growth rate (RGR_r). In contrast, *P. rugelii* showed only a small increase in reproductive biomass, and the correlation with vegetative biomass was weak.

Early protection against herbivores may result in many benefits, especially if the protected tissue has a high potential to gather additional resources in the future. Resources captured early are worth much more to the plant than similar quantities captured later, if these early-captured resources are allocated to organs, such as leaves, that can collect further resources. The analogy of this situation with compound interest in economics is obvious. However, Lerdau (1992) argues that economic models of investment do not work well for plants because there is no risk-free environment and economic models of compound interest require a risk-free setting. Economic analysis assumes that there is a trade-off among various sinks, and that resources are permanently allocated to these sinks. It is now clear that this may not be the case in all plants or all environments. Also, the allocation of resources such as minerals and proteins, independent of carbon and mass, are possible. Therefore, there is a need for other kinds of models to do a complete evolutionary analysis of allocation.

Whereas much attention has been given to allocation to shoots, including leaves, flowers, and fruits, much less attention has been given to roots and other underground parts. This is understandable because of the greater difficulty in studying roots in general. New techniques that allow a more accurate assessment of root growth and architecture (e.g., fiber optics, video imaging) are aiding biologists in their study of allocation. Whereas leaves of limited numbers of species, especially in wet environments, interact directly with other organisms such as algae and bryophytes (epiphylls), roots in the soil interact with a variety of soil microorganisms (fungi in mycorrhizal associations, nitrogen fixers, other bacteria, etc.). These organisms can greatly aid root function, in terms of both the availability and uptake of ions, and of water. It stands to reason then that roots must supply these organisms with lots of energy-containing compounds to sustain their growth. Measurements of standing roots biomass, therefore, may greatly underestimate the actual allocation to belowground parts. The partitioning of ions and water absorption between fine and coarse roots, respectively, and the discovery that in some plants there is hydraulic lift, i.e., the uptake of water from lower depths and its release in the upper parts of a soil profile (Caldwell and Richards, 1989), add further challenges to the problem. Furthermore, like herbivory on aboveground plant parts, herbivory on roots can be substantial and variable between years and habitats, but we still know very little about the extent and variation of belowground herbivory.

Acclimation, which can mean the restoration of the allometric ratios between plant parts, can occur at different rates in various species and within the same species for different traits. Many plants adjust their allocation in response to a changing environment (trackers) and are said to acclimate. In a heterogeneous environment acclimation is assumed to be functionally adaptive. Environmental shortages and excesses may greatly change the allocation patterns (see Chiariello and Gulmon, 1991). Soil moisture content and light levels can greatly influence the relative allocation to roots and shoots. Generally, it is assumed that this acclimation occurs in species with slow growth rates and therefore slow organ turnover rates (Grime and Campbell, 1991; Thornley, 1991). Long-lived organs in a changing environment must acclimate to that environment to optimize their resource gain. In contrast, short-lived organs can be discarded in favor of newer organs suitable to the environment. These two kinds of responses are seen in understory plants in the forest when a canopy gap is suddenly created above them, drastically changing the light environment.

We need more information to answer the following important questions:

1. How fast can a plant adjust its belowground and aboveground activities when either the shoots or the roots are subjected to severe herbivory and become out of synchrony with each other?

2. Does this adjustment happen by increasing the specific activity of shoots (enhanced photosynthetic rate per unit of leaf) and roots (enhanced

specific absorption rate), or does it happen largely by rapid reallocation of mass, and other resources, to reestablish the appropriate balance? Alternatively, is the adjustment done by killing off (cutting support from) some parts? It is possible that all these strategies are employed by plants, therefore, we need to know which strategy or combination of strategies is employed by what species, habitats, environments, or ecosystems.

3. What controls root turnover? Is it driven by the rate of leaf turnover to keep the plant as a balanced system? Leaf longevity is related to self-shading, density, and identity of neighbors, optimal nitrogen use, construction cost versus benefits, etc. (see Bazzaz and Harper, 1977; Field, 1983; Reich *et al.*, 1992). Limited evidence from our CO_2 enrichment experiments shows that for yellow birch and red maple seedlings root "birth" and death rates scale to shoot birth and death rates (Berntson and Bazzaz, 1996a,b), despite the difference between the two species in response to CO_2 enrichment.

II. Controls on Carbohydrate Manufacture and Allocation: The Role of Nitrogen

Triose phosphate (TP) is the first stable product of photosynthetic carbon fixation in plants. Triose is synthesized into sucrose, which then is exported to other plant parts, depending on demand (Fig. 1). This allocation seems to be closely regulated at several points to ensure a steady supply of sucrose to various active sinks during growth (Geiger and Servaites, 1991). Some extra sucrose is converted to starch and is stored until needed, either for growth or for osmotic adjustment under water stress conditions. The presence of some starch can promote growth because it forms a supply of necessary carbohydrates during the day as well as night periods (Schulze and Schulze, 1995). However, in some species the accumulation of much starch in cells may cause deformation of chloroplasts and down-regulation of photosynthesis.

Sink strength has been shown to greatly influence resource uptake, manufacture, and allocation of sucrose, a major transportable carbohydrate in plants. Down-regulation of photosynthesis can occur due to an imbalance between carbohydrate quantity and plant sinks as in the case of some plants grown in a CO_2-rich environment (e.g., Sage, 1994). Alternatively, there can be enhancement of photosynthesis when sink strength is increased by herbivory (e.g., McNaughton, 1983a; Detling *et al.*, 1980). In some species, contrary to expectations, aboveground herbivory also increases carbon allocation to roots and root exudates (Holland *et al.*, 1996).

Optimal allocation of mass to balance root/shoot ratios forms the bases for much of the modeling of the balanced growth of the plant (e.g., Iwasa

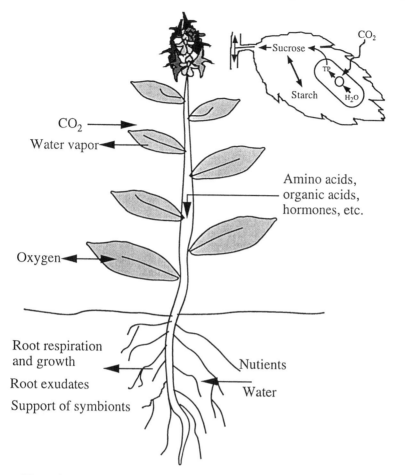

Figure 1   Uptake and processing of materials required for plant function.

and Roughgarden, 1984). However, resources may not be allocated independently from each other. For example, nitrogen seems to play a major role in the allocation of other plant resources. Ägren and Ingestad (1987) and Hilbert (1990) showed a strong linear relationship between plant nitrogen concentration and the fraction of mass allocated to leaves:

$$P \times S = dN/dt = aN,$$

where P is net photosynthesis, N is the nitrogen concentration, and a is a constant. Ägren (1985) also showed, by models, that relative growth rate is a linear function of nitrogen concentration in the plant (see also Hirose, 1986, 1988, Mooney and Winner, 1991). Cohen and Pastor (1996) developed a model of carbon balance and nitrogen uptake. Their model suggests that many of the observed correlations between plant traits (e.g., photosynthesis and nitrogen content) are best viewed as constraints on plant growth rather than cause and effect. It has also been shown that under limiting light conditions nitrogen accumulates as NO_3^-, which can be reallocated and used for growth if the external supply of nitrogen declines (Koch *et al.*, 1988). Also when the N supply is ample the relative growth rate (RGR) is directly proportional to photosynthetic rate, but when the N supply is limiting there is no effect of photosynthesis on growth (Fitchner *et al.*, 1995). When N is limiting starch can accumulate in cells as well (Schulze and Schulze, 1995). Heavy application of N fertilizer to the soil increased the requirement of the plant for K and Mg, inducing K deficiency in *Picea glauca* and *P. engelmannii* (van den Driessche and Ponsford, 1995). Thus, nitrogen nutrition and carbon gain and allocation are intimately related and must both be considered in allocation studies.

Ackerly (see Chapter 10, this volume) suggested that allocation to roots is related to plant growth in the following way:

$$df/dt = \text{RGR}(\partial - B_r/B_t),$$

where RGR is the relative growth rate, B_r/B_t is the root weight ratio, and ∂ is the fraction of new biomass invested in root growth. Relative growth rate is greatly influenced by leaf nitrogen content, and allocation to roots is strongly related to RGR.

The allometric relationship between roots and shoots in a changing environment is not constant but changes with ontogeny. This relationship has been found to be consistent with optimal allocation theory and is highly ontogenetically constrained (Gedroc *et al.*, 1996). It is, therefore, not surprising to find strong correlations between carbon allocation, growth, nitrogen content, and the partitioning of photosynthates between roots and shoots because all these activities are intimately related in plants. Nitrogen is needed for carbon gain (the manufacture of chlorophyll and *Rubisco*), carbon is need for nitrogen gain (root growth, support of symbionts, etc.), and both result in growth above- and belowground in a balanced way at some time scale. It is also not surprising to find weak correlation between CO_2 fixation and growth because of variation in allocation patterns. The differences between the two processes may be large. For example, in a broad survey, Körner *et al.* (1979) found up to 40-fold differences in growth and only 20-fold differences in photosynthetic rates in the same set of species.

III. Currency of Allocation and Costs of Construction: Keeping Track of Total Carbon Flux

Carbon has been considered the appropriate currency of allocation and cost (Reekie and Bazzaz, 1987a–c). It is assumed that carbon, in the form of sugars, is used to "purchase" other resources essential for plant function. Acquired carbon by plants via photosynthesis is also used to manufacture all other plant compounds. Poorter and Bergkotte (1992) identified the following groups as the major plant constituents: (1) lipids, (2) lignins, (3) soluble phenols (tannins and flavonoids), (4) organic N-containing compounds such as proteins and DNA, (5) total structural carbohydrates (e.g., cellulose), (6) nonstructural carbohydrates (e.g., starch, sugars), and (7) organic acids. The ratios of these classes change both with ontogeny and with the plant environment. Some of these materials show trade-off indicated by their negative correlations, whereas some show positive correlations such as that between proteins and minerals. These compounds also vary in their construction cost (the number of grams of glucose required to construct one gram of a given compound; Penning de Vries *et al.*, 1974). Lambers and Rychter (1989) estimate that the cost of lipids is 3.03, that of total nonstructural carbohydrates (TNC) is 1.09, and that of organic acids is 0.91 g glucose per gram. Plant parts also differ in cost of construction because they have different compositions (Williams *et al.*, 1987; Lambers and Rychter, 1989). Poorter (1994) estimates that the values are 1.5, 1.45, and 1.33 g glucose per gram for leaves, stems, and roots, respectively, and the differences among species are small. It must be remembered as well that cost can vary with the form of the unprocessed element. For example, it costs more to incorporate nitrogen into organic compounds within the plant if the nitrogen source in the soil is NO_3^- rather than NH_4^+, as the former must be converted to the latter before it can be metabolized. The conversion cost includes, among others, the manufacture of the enzyme nitrate reductase that catalyzes the reaction and also the reducing power from NADH, resulting in some competition between nitrogen reduction and photosynthetic dark CO_2 fixation.

There is also a relationship between construction cost and plant growth rate. If construction cost is high, the mass of constructed organisms must be lower for a fixed amount of carbon. Because of the consumption of carbohydrates in plant respiration, standing biomass (or carbon) measures underestimate the actual allocation to structures and functions. In contrast, keeping track of total carbon flux gives a realistic assessment of plant metabolism, as the overwhelming majority of the energy used in metabolism is obtained from photosynthetic carbon fixation.

IV. The Allocation of Resources Other Than Mass: Are N, P, K, Mg, and Others Allocated as a Fixed Proportion of Mass or Carbon?

Much of the theory of allocation to reproduction in plants has emphasized allocation of mass or carbon, and most studies do not consider the allocation of critical nutrients, e.g., N and P, to reproduction, which in some situations may greatly depart from the allocation of mass per se. Moreover, only a few studies have emphasized allocation in terms of resources after they are gathered and before they are permanently allocated to a given structure (see Lerdau and Gershenzon, Chapter 11, this volume). The allocation of plant resources that can be redeployed to various plant parts has not been adequately addressed, except for nitrogen, which is hypothesized to be allocated, or reallocated, to maximize whole plant carbon gain (see Field, 1983; Field and Mooney, 1986). Translocation of nitrogen from older, lower, more shaded leaves to young exposed leaves in the canopy occurs in many fast growing plants (Hirose and Werger, 1987a,b; Hirose *et al.*, 1996) and is in accordance with the optimal allocation hypothesis. Nevertheless, significant amounts of elements remain in tissues at the time of their senescence, especially in trees (Waring and Schlesinger, 1985). Resorption proficiency varies greatly among species for both nitrogen and phosphorus (Killingbeck, 1996). The cost of this loss of material to the plant is not well understood. We need answers to the following questions:

1. How much, if any, of the total plant mass, elemental levels, and manufactured resources are lost when plants reproduce or when they senesce? Why should the plant not convert the majority of these resources to productive structures?

2. Other than the general trend that gymnosperms are less proficient than angiosperms in resorption, are there within-group life history correlates with resorption proficiency?

3. Is the loss (if any) equal for all resources, or are there certain resources that are more conserved by the plant relative to other resources? Does that correlate strongly with mobility of the resource?

4. Do the ratios of resources in the plant change at the time of reproduction?

5. How much do the ratios of nutrients to mass and nutrients to each other change through ontogeny? And what flexibility exists in a population or a species for these ratios?

6. What is the minimal ratio allowable in a given stage of the life cycle of the individual before plant activity is drastically reduced?

7. Is there reallocation of defensive chemicals from vegetative structures to reproductive structures (flowers, fruits, and seeds)?

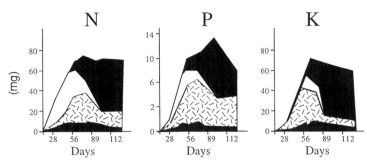

Figure 2 Allocation of N, P, and K to various plant parts in an annual plant. From top to bottom, fruits, leaves, stems, roots.

The limited available evidence suggests that the quantities of resources allocated from vegetative to reproductive structures are not trivial. There can also be loss of much material between maximum vegetative production and maximum reproduction.

The annual plant *Abutilon theophrasti* loses 31% of its nitrogen, 23% of its phosphorus, 12% of its potassium, and 27% of its calcium between peak growth and the time when reproduction is completed. *Datura stramonium* grown under the same conditions loses 30% of its nitrogen (similar to *Abutilon*) but 68% of its phosphorus, 30% of its potassium, and 55% of its calcium, which is much more than *Abutilon* (Benner and Bazzaz, 1987; Fig. 2). These changes also greatly modify the ratio of these resources (Table I). Surprisingly, this study also showed the importance of stems in the storage of mass and nutrients during growth and their redeployment during reproduction. Stems of *Abutilon* redeploy 31% of their mass, 60% of their nitrogen, 78% of their phosphorus, and 14% of their potassium to reproduction while *Datura* stems redeploy, respectively, 75, 60, 50, and

Table I Ratios of Mass, N, P, and K at Peak Standing Biomass and at End of Growth Period in *Abutilon*[a]

Ratio	Peak vegetative mass	End of growth
N/mass	106	82
P/mass	166	146
K/mass	90	89
N/P	63	56
N/K	120	92
P/K	186	164

[a] Mass and nutrient level data from Benner and Bazzaz (1985).

24% of their mass, nitrogen, phosphorus, and potassium to reproduction. Of the total resources accumulated, *Abutilon* allocates 45% of its mass to reproduction while *Datura* allocates up to 75%.

V. Physiological and Demographic Costs: Are They Separable?

Allocation of resources to growth and to defense of vegetative parts ensures the presence of some plant mass, part of which is later allocated to reproduction. Assessing the cost of reproduction and defense in plants has been considered at two levels: demographic cost, the reduction in survivorship of the reproducer and its future reproduction, and physiological cost, reduction in vegetative growth which indirectly influences survivorship and future reproduction, particularly in plants with a very strong correlation between size and reproduction. In many situations the two approaches are independently treated in experiments and models. In nature, however, they are intimately related. Physiological costs have demographic consequences. In a population, the growth of individuals has consequences to flowering, fruiting, and seed maturation and fecundity—a mixture of physiological and demographic parameters. Because of the great difficulty in assessing the allocation to long-term fitness in iteroparous, long-lived plants, physiologically based allocation has been advanced. The assumption is made that growth directly indicates long-term fitness (see Gershenzon, 1994). However, analyses have also shown that even physiological costs of reproduction may not be straightforward. For example, it has been shown that the degree of reduction in vegetative growth by reproduction can differ even among genotypes of the same species and in different environment (e.g., Jurik, 1985; Reekie and Bazzaz, 1987a,b). Reekie and Bazzaz (1992) found substantial variation in the cost of reproduction among several genotypes of *Plantago major* and *P. rugelii*. They also found that cost of reproduction in *Agropyron repens* was quite variable, depending on genotype, light level, and nitrogen availability (Reekie and Bazzaz 1987a,b; Fig. 3). Surprisingly, however, only a few tests of this variability in the field have been reported. Horwitz and Schemske (1988) found that reproduction did not measurably reduce growth or survivorship in a tropical herb. Thus, it remains to be seen how common is this type of response and how large must the difference be to be meaningful in the field. Are there specific life history correlates with this kind of response? Why do some genotypes and species have a small cost while others have a large cost?

I suggest that species that have little or no measurable cost of reproduction to growth may have the following attributes, assuming that other resources, especially nitrogen, are not limiting:

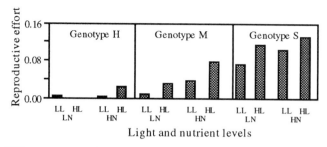

Figure 3 Differences among three genotypes and two light and two nutrient levels in reproductive effort in *Agropyron repens.*

1. Reproductive structures are green and photosynthetic. Therefore, they are able to supply part of their carbon need for reproduction, as has been shown in several species.

2. The plants are able to substantially increase their leaf area ratio (LAR, total leaf area divided by total plant weight) at the time of reproduction.

3. Experience substantially increases sink strength during reproduction, which has been shown to increase leaf photosynthetic rate as well.

Reproduction usually has a demographic cost, and in many plants may be a major cause of death of an individual (Harper, 1977). Therefore, there is usually a trade-off between reproduction and survival. Many models have been developed to address this relationship (see Caswell, 1989). One such model relates trade-offs as follows:

$$\text{Max } Vf = \int e^{rt} \, l_t b_t \, dt,$$

where r represents per capita increase, l_t is survival to age t, b_t is fecundity at age t, and t is age.

We investigated the cost of reproduction (the amount of lost vegetative growth per gram capsule produced) in genotypes of *Plantago major* and *P. rugelii* (Reekie and Bazzaz, 1992) and found good correlation between cost and the timing of the onset of reproduction. We found that the timing and extent of reproduction are related to differences between the two species in the effect of reproduction on growth. *Plantago rugelii* reproduced to a lesser extent than *P. major* because the cost per gram of capsule produced, in terms of reduced vegetative biomass, increased with reproductive output for the former species but not in the latter. Similarly, *P. major* reproduced earlier than *P. rugelii* because cost per gram of capsule increased with plant size but not in *P. rugelii*. Also, *P. rugelii* has to attain a larger mass before it reproduces. Minimum size (mass) for reproduction was

1.0 g for *P. major* and 3.6 g for *P. rugelii*. Thus, time of reproduction and cost of reproduction have both physiological and demographic costs.

VI. Allocation and Resource Congruency

Required resources for plant growth may not become available simultaneously. In some situations this incongruent resource availability may cause reduced growth and shifts in allocation. Some plants take up and store available resources (such as water and nutrients) until other resources become available and then process them together (see Bazzaz, 1996). In this case there must be costs of storage and protection, at least in the short term. For example, in forest gaps individuals located on the west side receive ample light in the morning hours when CO_2 levels are high, tissue water potential is less negative, etc. In this case, because these critical resources are available congruently, the plant can photosynthesize at a high rate (Bazzaz and Wayne, 1994). In contrast, plants located on the east side of a gap receive full sun only in the afternoon, when CO_2 levels are no longer high and plant tissue water potential has become more negative. In this case, resource processing can be greatly reduced (Wayne and Bazzaz, 1993a,b). How this *capacitance* (the ability of a plant to hold and protect resources until other essential resources become available; Bazzaz, 1996) influences the allocation of other resources is unknown. How do plants in different locations in a gap allocate resources to above- and belowground parts? Do plants in the northern sections of a gap in northern latitudes have greater allocation to roots, and how do plants allocate nitrogen between chlorophyll and the enzyme Rubisco in different locations in gaps? What about plants in the understory that experience intense sunflecks alternating with periods of low light levels?

VII. Switching from Vegetative Growth to Reproduction: Size, Mass, or Age?

Allocation of resources to reproduction in many plant species begins only after plants attain a certain mass, size, or age. This is especially evident in long-lived trees. It is generally assumed that after the attainment of this minimum size (mass) the relationship between size and reproductive output is positive and linear (e.g., Hartnett, 1990; Aarssen and Taylor, 1992, Méndez and Obeso, 1993). In fact, some authors feel that reasonable estimates of fitness can be obtained from size-classified matrix models (Caswell, 1989). However, in some species there is significant genetic variation and phenotypic plasticity in this relationship (e.g., Schmid and Weiner, 1993; Clauss and Aarssen, 1994). Moreover, the strength of this relationship may vary considerably among species and among the same species in different environments. For example, a high CO_2 environment seems to greatly reduce the strength of this relationship (Ackerly and Bazzaz, 1995). In another study with nine annuals (three species in each of three genera), we found no relationship between vegetative growth and reproduction in response to elevated CO_2 levels. Although hormonal imbalances (e.g., the increased production of ethylene and the accumulation of sugars in plant cells) may be involved, the actual causes of this relaxation or abolition of the relationship is not known, and its evolutionary consequences have not been fully addressed (see Farnsworth and Bazzaz, 1995).

It has been also assumed that in many plants attainment of a certain size rather than age is the critical factor for reproduction (e.g., Solbrig, 1981; Werner, 1979). In some plants minimum size to reproduction can differ with age. Mass and size, especially for perennials, while generally correlated within species, may have completely different relationships for different species or different environments. For example, in the herbs *Solidago* and *Aster,* a minimum size (mass) threshold for sexual reproduction is required (Schmid *et al.,* 1995). However, for clonal expansion, which is critical for occupation of more habitats, and for habitat choice (see Bazzaz, 1991), there appears to be no minimum mass for this mode of reproduction (Fig. 4).

Unless mass (which is usually measured as dry weight) and size (volume) are proved to be strongly and positively correlated, the relationship between size and onset of reproduction will remain fuzzy. As an example, we found significant correlation between growth of the target plant and the percentage of light intercepted by its neighbors (a functional parameter) rather than their leaf area or total mass (Tremmel and Bazzaz, 1993). We also found that the identity of neighbors can modify target allocation of its

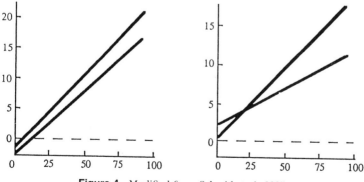

Figure 4 Modified from Schmid *et al.,* 1995.

mass. Metamer mass was more variable than its length or area suggesting that allocation flexibility allowed the plant to maximize size at the expense of mass (Tremmel and Bazzaz, 1995; Fig. 5). Species of the same community differ in their light transmission and therefore their influence on their neighbors (e.g., Pacala *et al.*, 1993). In natural populations plants occur in a wide range of densities, and the degree of crowding changes with the growth of the vegetation. There is little doubt that density and the architecture and activities of neighbors modify the allocation of one another. These intersections form the mechanistic bases for competition among adjacent plants.

More critically, it is the activity rate rather than mass or size that is more critical for understanding allocation. Equal leaf areas (or mass) can have vastly different photosynthetic rates and therefore will contribute different amounts to whole-plant carbon gain. Similarly, roots of the same mass, or length, may have great differences in their ion uptake capacity if they have different ion-specific absorption rates. Size, mass, resource gain competence, and age relationship to reproduction need to be understood further for a better assessment of the true costs and benefits.

The process of reproduction in plants entails a major shift in allocation of resources because allocation to vegetative growth, especially to leaves, is beneficial to plants in terms of further gain of carbon (Harper, 1985). Theoretically, an annual plant with a constant relative growth rate and a fixed growing season length should completely switch from vegetative growth to reproductive growth because relative growth rate decreases with time and the probability of death increases. Switching early can maximize seed sets (Cohen, 1976) unless delayed reproduction leads to an enhanced

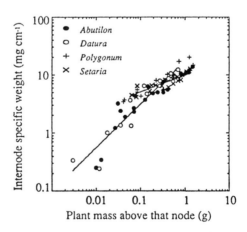

Figure 5 Relationship between plant mass above a node and internode-specific mass.

seed set (Chiariello and Roughgarden, 1984; see Bazzaz *et al.,* 1987, for a discussion). However, the switch in many plant species is gradual rather than abrupt, and plants do differ in their mixed allocation phase (King and Roughgarden, 1982) where both vegetative growth and reproduction take place simultaneously. Also, because phenological events in plants are dependent on environmental variation, the length of this mixed allocation phase and the switching to a completely reproductive phase of the life cycle are also variable and depend on environmental circumstances. For example, some plant species switch to reproduction during their first growing season (behave as annuals) where resources are plentiful but wait until the second or third years (behave as biennials or triannuals) when resources are in short supply or when the time to acquire these resources is limiting, for example, in a shorter than usual growing season (e.g., Reinartz, 1984; de Jong, 1986).

Species vary in their ability to adjust their within-organ allocation as well. McConnaughay and Bazzaz (1992) found that when surrounded by inert objects, simulating the physical presence of neighbors, leaves that were able to escape this restriction had much longer petioles and much smaller leaf blades than those that remained trapped. This trade-off between petiole length and blade size probably increased the potential for carbon gain (exposed blades are more likely to attain a higher photosynthetic rate than those that remain shaded in the canopy). It is not known whether plants in nutrient-limited environments allocate more to total root length or mass of fine roots. The ability to shift within-organ allocation is a component of the plant flexibility in variable environments. There are trade-offs between lengths of petioles and blade size or mass, and a cost–benefit analysis of reduction in blade size and an increased per unit area photosynthetic rates is needed to understand the net carbon gain of this response.

Allocation within seeds can vary as well. Total seed weight can be allocated to the embryonic axis, to the endosperm, when present, and to the seed coats. Each of these components can differ in mass and in relation to each other depending on species and habitat. Seed coats can be thin or thick, accounting for a relatively small percentage of total seed mass or a relatively large percentage of total seed mass (Fig. 6). Seed coats can be viewed as

Figure 6   Differences in seed coat mass between seeds having the same mass.

protective against the physical environment and against seed predators. Are they then part of reproductive allocation or allocation to defense? Do seed coats differ from embryos in their chemical composition? Are the differences in nitrogen to carbon content large enough in some seeds to warrant separate consideration of allocation?

VIII. Resource Allocation to Reproduction: Quantity and Quality

Although it is easy to estimate short-term reproductive allocation (resources contained in reproductive structures in one growing season), it is much more difficult to estimate long-term reproductive effort for the entire life cycle of a semelparous, long-lived plant (Bazzaz and Reekie, 1985). Lifelong reproductive effort and its direct and indirect costs (see Reekie and Bazzaz, 1987a–c), which are the most relevant to allocation theory, are estimable only if the general shape of the reproductive time curves for various species is well-established (Bazzaz and Ackerly, 1992).

The proportion of standing resources contained in reproductive structures, relative to the total resources obtained and manufactured in the plant, can vastly underestimate the cost of reproduction. Based on data in Whittaker (1966) on the growth of deciduous trees, reproductive allocation (RA) can be 60 times bigger when expressed as incremental increases in tree biomass than when expressed on the basis of tree total biomass. Fine roots, which are responsible for much of the uptake of nutrients from the soil medium, are almost always underestimated because they are lost in excavating and washing. Although this loss is small relative to the large mass of coarser roots, it may cause a large bias in the assessment of nutrient uptake, which is determined largely by fine roots.

Allocation to reproduction, if measured as the percentage of standing biomass in reproductive structures, may not be an exact measure of allocation of resources by the plant into all the processes that lead to the production of seed. In its totality the latter allocation represents the actual effort invested by the plant for seed production. Thus, a distinction must be made between RA and reproductive effort (RE). Unless the two measures are closely related, only RA is useful for addressing the evolutionary history of the species. Bazzaz and Ackerly (1992) discuss in detail factors that can decouple these two measures. Based on Reekie and Bazzaz (1987c) they propose the following formulation for assessing reproductive effort:

$$\text{RE} = \frac{(R_r + R_u + S_r + A_r) - P_r}{(T_r + S_v + A_v - P_r)},$$

where R_r represents the reproductive pool, R_u is vegetative biomass attributable to reproduction, S_r represents structural losses from reproductive organs, A_r represents atmospheric losses from reproductive organs, P_r is enhancement of total resource supply due to reproduction, T_r is the total standing pool, S_v represents structural losses from vegetative organs, and A_v represents atmospheric losses from vegetative organs. Whereas some of these parameters are not easy to estimate, they are required for any complete analysis of true cost and benefits.

In some plants allocation to reproductive organs (e.g., flowers) may exceed the ability of the plants to mature all because of resource limitations. Overproduction of flowers and subsequent abortion has been observed in many plants (e.g., Lee and Bazzaz, 1982a,b; Marshall and Ellstrand, 1988). In these cases abortion can be a mechanism for adjusting reproductive output to the level of resources available in the particular habitat (Lloyd, 1980; Stephenson, 1981). In a variable environment this strategy may be beneficial despite the fact that plants do lose some resources invested in, but not translocated from, dying flowers, a "bet hedging" strategy. It also may involve some evolutionary advantage rather than proximal loss in terms of mate choice by selective abortion of flowers and fruits sired by less fit fathers (Willson, 1983; Lee, 1988). This strategy represents short-term physiological loss for long-term evolutionary gain.

Many studies on growth and productivity at the community and ecosystem levels assume that vegetative growth is strongly coupled with reproductive allocation. Therefore, estimations of standing plant mass are used to deduce future ecosystem structure and function. However, from an evolutionary point of view, fecundity is more critical than mass per se. The degree of coupling between vegetative growth and reproduction may vary in strength, and there are environmental circumstances where the two are only weakly related (Ackerly and Bazzaz, 1995). Chiariello and Gulmon (1991) proposed that the degree of coupling is influenced by the morphology of the inflorescence. Flowers that develop in the axils of leaves are highly coupled, whereas terminal inflorescences are less coupled (Fig. 7). The assumption is made that leaves which subtend flowers supply them with much of the resources required for fruit maturation. This assumption, however, remains to be tested.

An important issue, which has not been adequately considered in the analysis of allocation to reproduction, is the quality of the progeny. It has been shown that seeds differ in chemical composition from vegetative parts, and there can be large differences among seeds of individuals grown in different environments. *Abutilon* individuals grown in a range of soil nutrients have been shown to contain significantly different levels of nitrogen. Seeds with high nitrogen content are more competitive than seeds with low nitrogen content (Parrish and Bazzaz, 1985a). Furthermore, *Ambrosia artemisiifolia* and *Abutilon theophrasti*, which cooccur in the field when grown

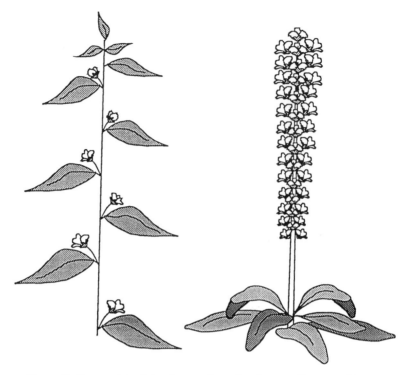

Figure 7   Two models of the degree of coupling between flowers and leaves.

under identical nutrient conditions, produced seeds with vastly different nutrient content (Fig. 8). In nine annual species seed quality and fitness, in terms of viability and germination, were not always directly correlated with the quantity of seeds produced (Farnsworth and Bazzaz, 1995). Elevated CO_2 had significant influence on seed germination in the three species of *Ipomoea* but little influence in seed germination in the three species of *Polygonum*. This study found no relationships between vegetative growth and reproductive response to the treatments. In *Polygonum persicaria* seed quality can differ depending on the environment of the mother (Sultan, 1996). Differences in seed quality can have significant influence on competitive ability and the contribution to future generations. Quality may be more important than quantity for the future of a genotype.

IX. Clonal Plants and Allocation: Are There Trade-offs between Sexual and Asexual Reproduction?

Clonal plants are common in many habitats, including the understory of forests and mid-successional habitats. A challenge to allocation theory

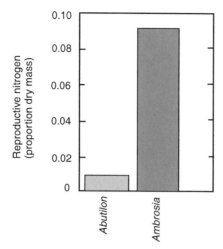

Figure 8   Nitrogen content of *Abutilon* and *Ambrosia* seeds grown under the same conditions.

is presented by these plants because they produce new individuals both sexually through seed and asexually by elaborating eventually independent individuals. The relationship between these two forms of reproduction is not well understood. Are there one-to-one trade-offs between these two forms? Is the allocation of more resources to one mode or the other dictated by the environment? Is the degree of patchiness in the spatial distribution of critical resources for plant growth the overriding factor in directing allocation? Can the presence of competitively superior neighbors cause a shift of more resources to sexual reproduction (and escape) rather than to asexual reproduction? Is one mode of reproduction more costly than the other in terms of reduction in growth? The available literature sheds limited light on these issues despite their great importance to allocation theory.

When they invade a habitat, clonal plants usually start from seeds and in subsequent years grow and reproduce (spread) asexually from rhizome or stolon buds. It is quite uncommon to find seed-derived individuals after the initial phase of invasion of old fields (Bazzaz, 1996). Under field conditions individuals of the common *Solidago* derived from rhizomes are not distinguishable, in terms of either size or phenology, from individuals derived from rhizomes. However, these forms differ in their allocation patterns and in timing of life history events. Those derived from rhizomes flower and fruit at a smaller mass than those derived from seeds, which, in many situations, remain completely vegetative (Fig. 9). Connections between daughter ramets and their usually larger parent may remain for a long time in some species or only for a short time in others. The kind and the quantity of resources translocated to daughter ramets (and vice versa) are not fully known, but they are likely to involve sugars, amino

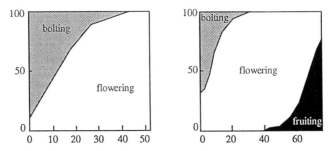

Figure 9   Differences in phenology between seed-derived (left) and rhizome-derived (right) plants of *Solidago canadensis.*

acids, water, and nutrients (see Caldwell, 1994; Hutchings and de Kroon, 1994). The length of the period of support in different clonal plants may depend on the shape of the relationship between size (mass) and fitness. Cardaco and Kelly (1991) developed models in which they predict that a long connection is favored if the translocation of assimilates causes a large increase in the fitness of the daughter ramets. However, if this translocation does not lead to such an increase, then connections are severed and daughter ramets quickly become independent.

Although our knowledge is still limited in this area, it is reasonable to assume that the rate of transport and the identity of the transported materials will depend on sink strength for that material. For example, when soil moisture is heterogeneous, water translocation between connected ramets becomes unidirectional. de Kroon *et al.* (1996) show that up to 60% of water taken by ramets in wet soil is transported toward ramets of the same genet in dry soil. It is quite likely that a lager "mother" ramet should always translocate resources to its smaller "daughter" ramet. However, a small mother ramet, if it can, should keep resources to herself. It is also reasonable to assume that different resources can move fairly independently of each other following concentration or demand gradients. The allocation of limiting resources to connected offspring enlarges the foraging domain of the entire genet (Bazzaz, 1984, Pitelka and Ashman, 1985) and may result in the integration of patches that are heterogeneous in resources occupied by a single genet (Hartnett and Bazzaz, 1985).

The distinction between asexual reproduction and growth per se can confuse the understanding of allocation. The production of new ramets can be thought of as growth leading to clonal expansion. However, when the connections between ramets are severed, each independent ramet becomes a separate individual. Does severing rhizome connections distinguish between growth and reproduction? Are the mass and nutrients allocated to growth the same as the material allocated to reproduction?

The situation in dioecious clonal plants may even be more complicated. For example, in *Antennaria parlinii*, a common herb in some successional habitats, there are sexual seeds, apomictic seeds, and ramets developed from stolons which are severed after some time. We found that the allocation to ramet production was different among male, female, and apomictic plants (Michaels and Bazzaz, 1986; Fig. 10). Apomictic individuals allocated least to stolon production and had the highest frequency of mortality, whereas males allocated most to stolon production and had the lowest mortality. Under the same conditions sexual populations produced more total mass than did the apomicts. However, in experiments on controlled light and nutrient gradients, the reproductive mass of the apomicts was greater than that of the sexual populations on all resource levels. Also, percent allocation to reproduction in apomictic individuals increased dramatically with an increase in resource level, but the response of the sexual individuals was stable on these gradients (Fig. 11). Allocation in the apomictic plants, which are more common in disturbed ground, make them suited for these habitats. The plants are highly plastic in allocation and opportunistic in reproduction. Similarly, allocation in the sexual individuals (which are more common in less disturbed habitats) promotes competitive ability with emphasis on vegetative growth over seed production (Michaels and Bazzaz, 1989).

To summarize, it is now apparent that activities and functions are not always mutually exclusive in the allocation to growth (e.g., N initially can be translocated to upper unshaded leaves and later to reproduction). All

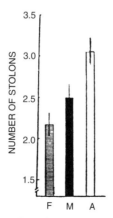

Figure 10   Seed production among females (F), males (M), and apomictic individuals (A) of *Antennaria*.

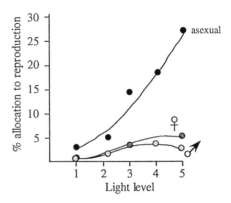

Figure 11   Allocation to reproduction in males, females, and asexual plants in *Antennaria* on a light gradient.

the 18 kg of water needed for inflorescence development in *Agave* can be supplied by reallocating water from leaves (Nobel, 1977). Different plant organs may perform different primary plant functions (leaves for photosynthesis, roots for water and nutrient uptake, etc.) but, for the most part, additionally carry on other plant functions. Some organic compounds could be broken down to generate building blocks for materials that can be used for other purposes, such as defensive chemicals. Resources are not necessarily in constant supply from the environment, and many plants have developed mechanisms to obtain what is most limiting or balance their uptake (acquisition) rate such that several resources become equally limiting (Bloom, 1985, Chapin *et al.*, 1987). Photosynthesis of reproductive parts can pay for part of the carbon cost of building reproductive biomass (Bazzaz *et al.*, 1979; Reekie and Bazzaz, 1987a; Bazzaz and Reekie, 1985).

X. Allocation to Defensive Chemicals

Plants have been subjected to herbivores from the beginning of their evolutionary history. Indeed, insect galls have been discovered in fossil plant material from 300 million years ago (Labandeira and Phillips, 1996). Many plants have developed mechanisms that prevent their demise by using morphological, behavioral, and chemical defenses. Defensive chemicals, which are usually classified as carbon-based or nitrogen-containing compounds, are thought to be by-products of primary metabolism. There is now little doubt that at least some of these compounds are specifically manufactured by plants for the purpose of their defense against herbivores (see Fritz and Simms, 1992).

Like allocation to reproduction, the theory of allocation to defense in plants rests on principles of cost/benefits and trade-offs between various structures and functions shaped by natural selection (Rhoades, 1979; McKey, 1979; Coley *et al.*, 1985; Rosenthal and Janzen, 1979; Zangerl and Bazzaz, 1992). Many of the analyses of the cost of defense make general assumptions about the cost of the manufacture of defensive chemicals. With advances in plant biochemistry, it is now possible to accurately calculate the cost of biosynthesizing and storing defensive chemicals. These kinds of analyses are producing surprising results. For example, Lerdau and Gershenzon (Chapter 11, this volume) present a thorough analysis of the cost of monoterpenes, defensive chemicals in the genus *Pinus*. They show that the cost of storing these compounds in specialized structures outweighs the cost of their manufacturing. In addition to these costs, they identify transport, maintenance, and forgone resource capture (opportunity cost) as the total defensive cost incurred by the plant. Is this the case in other defensive chemicals? Under what circumstances is storage more costly than manufacturing defensive chemicals?

Optimal defense theory assumes the following; (a) defense has cost, (b) not all plant parts are equally valuable to the plant and therefore are not equally defended, and (c) the cost of defensive chemicals can be substantial not only in terms of manufacture and storage but also in terms of opportunity cost, as energy allocated to defense cannot be simultaneously used to obtain more energy. The cost of various defensive chemicals also differs among themselves. For example, alkaloids cost more than tannins (e.g., Chew and Rodman, 1979; Gulmon and Mooney, 1986; Fritz and Simms, 1992). Also, most herbivory models assume that herbivory is predictable. However, the limited tests of the assumption produce mixed results (Karban and Adler, 1996).

In most cases the level of herbivory in the field is negatively correlated with the level of secondary compounds in the plant (e.g., Coley, 1983). However, calculating the exact cost of chemical defenses and their benefit to plants has been difficult to achieve. Zangerl and Bazzaz (1992) show negative correlation between defense and fecundity for plants in general. However, these analyses will remain incomplete unless the return on this cost (in terms of growth and fecundity) is measured with and without these defenses. The use of susceptible and resistance genotypes is aiding our understanding of this situation. However, the techniques of molecular biology, by which the gene(s) that controls production of chemicals can be deleted, have not been fully utilized in this analysis.

⋅ Inducible defenses (chemicals whose concentration greatly increases on attack by herbivores) make up a cost-saving strategy, as they are manufactured only when they are needed and there is no cost associated with their storage. This strategy, however, may be disadvantageous in habitats where there is a high probability of attack (Rosenthal and Janzen, 1979) or where plants are bound to be found (Feeney, 1976; Dirzo, 1984). Plants in such habitats must depend on chemicals that are present throughout their life cycle or at stages where they are most likely to be attacked by herbivores. Zangerl and Bazzaz (1992) proposed a graphical model that relates the proportional reliance on inducible and constitutive defensive chemicals to the probability of attack on the plant (Fig. 12).

There are several models dealing with cost and benefits of plant defense. Based on the data of Coley *et al.* (1985), Basey and Jenkins (1993) developed a model of investment in plant defense. The model has the following formulation:

$$R = dC/dt = CG(1 - kD^{\alpha}) - (H - mD^{\beta}),$$

where R represents realized growth, C is plant biomass at time 0, G is the maximum inherent growth rate in the absence of herbivores, kD is the proportion of growth reduction due to investment in defense, D is the investment in defense, $H - mD^{\beta}$ is the reduction in realized growth due to herbivory, H is the potential for herbivore pressure without the presence of defense, and $m, \beta, k,$ and α are constants. They concluded from this model that plants with high inherent growth rates should be without immobile "quantitative" defenses, whereas plants with slow growth rates should have a high level of these defenses, with no intermediates between these two types. Loehle and LeBlanc (1996) presents evidence for the presence of these two extreme types of defenses from an extensive analysis of several North American broad-leaved trees and several North American pines.

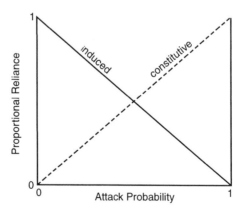

Figure 12   Trade offs between constitutive and induced defenses in plants in relation to the probability of attack.

Adler and Karban (1994) suggest a third possibility of defense. They call it the "moving target." Plants employing this strategy respond to herbivory by altering their phenotype. Their model shows that constitutive defenses are favored when herbivory is relatively constant and the cost of defense is high. Inducible defenses are favored when herbivory varies and the cost of defense is not too high. Moving target defense occurs when a plant phenotype is effective against one herbivore but is not effective against another herbivore. Zangerl and Bazzaz (1992) proposed a simple cost–benefit analysis of defense of leaves. Their model is based on a demographic technique developed by Bazzaz and Harper (1977) and incorporates the indirect cost of both defense and herbivory. Thus, leaf production rate is written

$$dL/dt = L \times H \times (B - C) + L \times H$$

where L is the number of leaves on the plant, B is the leaf specific birth rate (leaves/leaf/week), C is the number of leaf equivalents allocated to defense, and H is the proportion of leaf equivalents surviving herbivory owing to defense. In this equation the first term estimates the indirect cost of herbivory and defense (in terms of reduced production of leaves), and the second term estimates the direct cost of herbivory (in terms of leaf area lost to herbivores). Simulations with this model, using data of Coley (1986) on *Cecropia paltata,* an early successional, tropical tree, show that allocating the equivalent of 6% of leaf biomass to defense causes a 33% reduction in growth after 18 months. Coley (1986) shows that the level of herbivory on less defended genotypes (low tannin) is four times the level of well defended ones (twice as much tannin). Our model shows that the level of allocation to double tannin content far outweighs its cost in terms of carbon gain (see Zangerl and Bazzaz, 1992, for further details).

Large-scale ecological correlates for the allocation of defensive chemicals show that high allocation is associated with evergreenness (Janzen, 1984) and resource-limited environments occupied by plants having slow growth rates (Mooney and Gulmon, 1982; Coley, 1983; Bryant *et al.,* 1983). Reproductive allocation, which should have the opposite correlations (if there are trade-offs between defense and reproduction), has not been considered in this way except for the general notion that *r*-selected species (usually fast growers) allocate more to reproduction than *K*-selected species, which are usually slow growing.

While allocation to defensive activities have been considered for both herbivores and, to a lesser extent, pathogens in wild plants, there is much less consideration of allocation to symbiotic association, such as N-fixing bacteria in nodules of legumes and mycorrhizal associations, which seem to be present in most species and habitats and are crucial to plant establishment and growth. The limited available evidence suggests that allocation of resources (e.g., sugars) to these symbionts can be very large and may exceed allocation to defensive compounds. Therefore, in species that rely on these symbionts this allocation must be considered in the analysis of cost/benefits in plants.

Defensive allocation can be highly localized because not all plant parts are equally valuable to the plant. The loss of a particular organ such as a young leaf, a flower, or a fruit can have a large negative impact (reviewed by Rupp and Denno, 1983; Krischik and Denno, 1983). In some species defensive chemicals are allocated differentially within organs, for example, near leaf margins, seed coats, and the outside of fruits, areas exposed to attack by herbivores (e.g., Berenbaum and Zangerl, 1986). Within leaves, allocation of these chemicals can vary depending on the importance of various areas to the proper functioning of the entire organ (Fig. 13). For example, in *Pastinaca sativa* about half of the total leaf furanocoumarins are found in the veins which, if damaged, would greatly reduce the carbon-gain capability of the leaves in the part distal to the point of damage because that part will be deprived from water and nutrients. Seeds and fruits are rich in carbon compounds and nutrients and are strongly linked to fecundity. It is expected, therefore, that they are highly defended. For example, the concentration of the defensive chemical furanocoumarin increases from flower bud to a fruit in *Pastinaca sativa* (Fig. 14). In dioecious plants, available evidence (e.g., Ägren, 1987; Elmqvist and Gardfjell, 1988; Nitao and Zangerl, 1986) suggests that males are damaged more than females. Thus, females incur a higher cost of reproduction as well as a higher cost

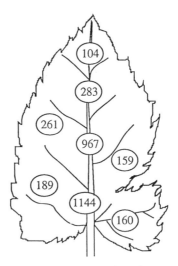

Figure 13   Distribution of furanocoumarin in various parts of *Pastinaca* leaves.

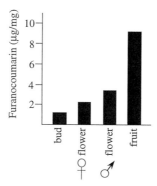

Figure 14   Furanocoumarin content in flower buds, flowers, and fruits of *Pastinaca.*

of defense. Only a complete cost–benefit analysis would show why females incur such a high cost.

XI. Allocation in a Globally Changing Environment

Global change involves, among other things, an increase in atmospheric carbon dioxide concentration, increased nitrogen deposition, and an increase in temperature (Vitousek, 1994; Houghton *et al.*, 1995). These factors have direct relevance to allocation of resources in plants. Increased plant growth under new environmental conditions leading to nitrogen dilution can have major consequences to allocation of other resources under these conditions because of the strong relationship between plant nitrogen content and allocation patterns previously discussed in this chapter.

Global change conditions influence the quantity, identity, and ratios of various resources needed for plant growth, reproduction, and defense. Elevated CO_2, nitrogen deposition, and elevated temperatures, by enhancing growth, may also accelerate competition among neighbors and change mass hierarchy in plants. They can have both physiological and demographic costs. Allocation of mass to reproduction can change under these conditions. Higher temperature can result in higher germination rates, faster initial growth rates, and modified survivorship. Altered mass hierarchy can result in a reduced effective population size (N_e) and lower percentages of plants flowering and producing seeds (Bazzaz and Morse, 1991). Furthermore, both temperature rise and elevated CO_2 change plant architectural allocation by shifting the relationship between total number of nodes, aboveground biomass, and the minimum size to flowering. Under conditions of an elevated CO_2 environment of the future, belowground allocation can increase in some species (Bazzaz, 1990; Rogers *et al.*, 1994). It can

Number of rhizomes   Mean rhi. len. (cm)

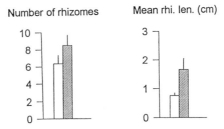

Figure 15   Influence of elevated CO_2 on the production of rhizomes and on rhizome length in *Solidago*. Ambient CO_2 (unshaded), elevated CO_2 (shaded). Chong and Bazzaz (unpublished).

enhance root turnover (Berntson and Bazzaz, 1996a) and increase root exudates (reviewed by Stulen and der Hertog, 1993; Rogers *et al.*, 1994). Increased root exudates and support of mycorrhizal fungi under elevated CO_2 conditions may lead to even greater errors in estimating allocation and understanding the fate of the carbon fixed in photosynthesis.

The clonal herb *Solidago canadensis* allocates more to new rhizomes and produces longer new rhizomes under conditions of elevated CO_2 (Fig. 15). This may lead to fast expansion of such species in the future. The influence of increased allocation to rhizome production on sexual reproduction and escape to new habitat is not known.

Many studies have consistently shown that tissues of plants exposed to elevated CO_2 environments have a high C/N ratio (Fig. 16). It is usually assumed that insect herbivores eat leaves for their nitrogen content (Matt-

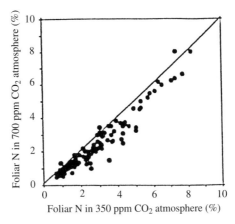

Figure 16   Foliar leaf nitrogen content is reduced when plants are grown in a high CO_2 environment. Modified from Traw *et al.*, 1996.

son, 1980). Analysis of data in some species has also shown that this nitrogen dilution in plants grown in an elevated CO_2 environment may not be due to a change in physiology such as increased nitrogen use efficiency, but rather to accelerated growth under a high CO_2 environment. When plants are compared at the same size rather than the same age, differences in nitrogen content between plants grown in ambient and high CO_2 environments disappear (Coleman *et al.*, 1993; Agren, 1994). The total nitrogen content and its concentration in leaves and other tissues subject to herbivory in the world of the future need a more thorough examination. High CO_2 and high nitrogen grown plants grew faster than low nitrogen, ambient CO_2 plants (Körner and Bazzaz, 1996). How might this change in nitrogen concentration and measured growth rate influence allocation? Do plants grown in a high CO_2 environment of the future allocate more to carbon-based defense as predicted by the carbon/nutrient hypothesis of Bryant *et al.* (1983)? Will the relationship between defense and reproductive allocation change? (See discussion in Fajer *et al.*, 1992).

There is only a limited knowledge of these issues. Fajer *et al.* (1992) found that in the herb *Plantago lanceolata* individuals grown under low nutrient conditions have high concentrations of carbon-based allelochemicals (acubin, catalpol, and verbascoside). However, individuals grown in elevated CO_2 environments had similar or lower concentrations of these compounds. In this experiment individuals grown in elevated CO_2 environments had greater reproductive biomass only under high nutrient conditions. In birch, however, we found a significant increase in tannin content in seedlings grown under high CO_2 environments (Traw *et al.*, 1996).

New sets of ecologically important questions about allocation arise in a changing climate. Global change involves, among other things, an increase in the level of two critical plant resources, CO_2 and nitrogen. These two intimately related resources are taken up usually by two different parts of the plant: leaves and roots, respectively. However, slow decomposition rates, competition between plant roots and soil microbes for nitrogen, and the possible change in the form of nitrogen deposition may modify the nitrogen (and the carbon) cycle in some ecosystems (see Vitousek, 1994). The limited number of studies do not show clear trends. We therefore may ask the following:

Will the allometric relationships between plant parts change?

Will there be any change in root turnover rate? Will it scale with the rate of leaf turnover rate and with root growth rate?

Will there be an increase in root exudates? Will their chemical identity change?

Will there be an increase in carbon-based defensive chemicals?

Will tissue toughness, water content, nitrogen concentration, and total nonstructural carbohydrates (TNC) change? How might these changes influence the behavior and the population biology of herbivores?

The above discussion of allocation to various activities, structures, and functions clearly indicates that we have learned a great deal over the last two decades about how plants defend themselves, and accomplish enough growth to allocate part of their mass to keep their lineage. This discussion also shows that some crucial questions about allocation strategies and tactics remain incompletely answered. Global change conditions add another set of questions that were not considered in the past. Therefore, the study of allocation will continue to be a rich and very exciting area of plant biology.

References

Aarssen, L. W., and Taylor, D. R. (1992). Fecundity allocation in herbaceous plants. *Oikos* **65,** 225–232.

Ackerly, D. D., and Bazzaz, F. A. (1995). Plant growth and reproduction along CO_2 gradients: Non-linear responses and implications for community change. *Global Change Biology* **1,** 199–207.

Adler, F. R., and Karban, R. (1994). Defended fortresses or moving targets? Another model of inducible defenses inspired by military metaphors. *Am. Nat.* **144,** 813–832.

Ägren, G. I. (1985). Theory for growth of plants derived from the nitrogen productivity concept. *Physiol. Plant.* **64,** 17–28.

Ägren, G. I. (1987). Intersexual difference in phenology and damage by herbivores and pathogens in dioecious *Rubus chamaemorus* L. *Oecologia* **72,** 161–169.

Ägren, G. I. (1994). The interaction between CO_2 and plant nutrition: Comments on a paper by Coleman, McConnaughay and Bazzaz. *Oecologia* **98,** 239–240.

Ägren, G. I., and Ingestad, T. (1987). Root:shoot ratio as a balance between nitrogen productivity and photosynthesis. *Plant Cell Environ.* **10,** 579–586.

Amthor, J. S. (1995). Higher plant respiration and its relationships to photosynthesis. *In* "Ecophysiology of Photosynthesis" (E.-D. Schulze and M. M. Caldwell, eds.), pp. 71–101. Springer-Verlag, Berlin.

Basey, J. M., and Jenkins, S. H. (1993). Production of chemical defenses in relation to plant growth rate. *Oikos* **68,** 323–328.

Bazzaz, F. A. (1984). Demographic consequences of plant physiological traits: Some case studies. *In* "Perspectives in Plant Population Ecology" (R. Dirzo and J. Sarukhan, eds.), pp. 324–346. Sinauer, Sunderland, Massachusetts.

Bazzaz, F. A. (1990). The response of natural ecosystems to the rising global CO_2 levels. *Ann. Rev. Ecol. Syst.* **21,** 167–196.

Bazzaz, F. A. (1991). Habitat selection in plants. *Am. Nat.* **137,** S116–S130.

Bazzaz, F. A. (1996). "Plants in Changing Environments: Linking Physiological, Population, and Community Ecology." Cambridge Univ. Press, Cambridge.

Bazzaz, F. A., and Ackerly, D. D. (1992). Reproductive allocation and reproductive effort in plants. *In* "Seeds: The Ecology of Regeneration in Plant Communities" (M. Fenner, ed.), pp. 1–26, C.A.B. International, Wallingford, Oxon, U.K.

Bazzaz, F. A., and Harper, J. L. (1977). Demographic analysis of the growth of *Linum usitatissimum*. *New Phytol.* **78,** 193–208.

Bazzaz, F. A., and Morse, S. R. (1991). Annual plants: Potential responses to multiple stresses. *In* "Integrated Response of Plants to Stress" (H. A. Mooney, W. E. Winner, and E. J. Pell, eds.), pp. 283–305. Academic Press, Orlando, Florida.

Bazzaz, F. A., and Reekie, E. G. (1985). The meaning and measurement of reproductive effort in plants. *In* "Studies on Plant Demography: A Festschrift for John L. Harper," pp. 373–387. Academic Press, London.

Bazzaz, F. A., and Wayne, P. M. (1994). Coping with environmental heterogeneity: The physiological ecology of tree seedling regeneration across the gap–understory continuum. *In* "Exploitation of Environmental Heterogeneity by Plants: Ecophysiological Processes Above- and Belowground," (M. M. Caldwell and R. W. Pearcy, eds.), pp. 349–390. Academic Press, San Diego.

Bazzaz, F. A., Carlson, R. W., and Harper, J. L. (1979). Contribution to reproductive effort by photosynthesis of flowers and fruits. *Nature (London)* **279**, 554–555.

Bazzaz, F. A., Chiariello, N., Coley, P. D., and Pitelka, L. (1987). The allocation of resources to reproduction and defense. *BioScience* **37**, 58–67.

Benner, B. L., and Bazzaz, F. A. (1985). Response of the annual *Abutilon theophrasti* medic. (Malvaceae) to timing of nutrient availability. *Am. J. Bot.* **72**, 320–323.

Benner, B. L., and Bazzaz, F. A. (1987). Effects of timing of nutrient addition on competition within and between two colonizing annual plants. *J. Ecol.* **75**, 229–245.

Berenbaum, M. R., and Zangerl, A. R. (1986). Variation in seed furanocoumarin content within the wild parsnip (*Pastinaca sativa*). *Phytochemistry*, **25**, 659–661.

Berntson, G. M., and Bazzaz, F. A. (1996a). Nitrogen cycling in microcosms of yellow birch exposed to elevated CO_2: Simultaneous positive and negative feedbacks. *Global Change Biology* (in press).

Berntson, G. M., and Bazzaz, F. A. (1996b). The allometry of root production and loss in seedlings of *Acer rubrum* (Aceraceae) and *Betula papyrifera* (Betulaceae): Implications for root dynamics in elevated CO_2. *Am. J. Bot.* **83**, 608–616.

Bloom, A. J. (1985). Wild and cultivated barleys show similar affinities for mineral nitrogen. *Oecologia* **65**, 555–557.

Bloom, A. J., Chapin III, F. S., and Mooney, H. A. (1985). Resource limitation in plants—An economic analogy. *Annu. Rev. Ecol. Syst.* **16**, 363–392.

Bryant, J. P., Chapin III, F. S., and Klein, D. R. (1983). Carbon/nutrient balance of boreal plants in relation to vertebrate herbivory. *Oikos* **40**, 357–368.

Caldwell, M. M. (1994). Exploiting nutrients in fertile soil microsites. *In* "Exploitation of Environmental Heterogeneity by Plants: Ecophysiological Processes Above- and Below-ground" (M. M. Caldwell and R. W. Pearcy, eds.), pp. 325–347. Academic Press, San Diego.

Caldwell, M. M., and Richards, J. H. (1989). Hydraulic lift: Water efflux from upper roots improves effectiveness of water uptake by deep roots. *Oecologia*, **79**, 1–5.

Caraco, T., and Kelly, C. K. (1991). On the adaptive value of physiological integration in clonal plants. *Ecology* **72**, 81–93.

Caswell, H. (1989). "Matrix Population Models." Sinauer, Sunderland, Massachusetts.

Chapin III, F. S. (1991a). Integrated responses of plants to stress. *BioScience* **41**, 29–36.

Chapin, III, F. S. (1991b). Effects of multiple environmental stresses on nutrient availability and use. *In* "Response of Plants to Multiple Stresses" (H. A. Mooney, W. E. Winner, and E. J. Pell, eds.), pp. 67–88. Academic Press, San Diego.

Chapin III, F. S., Bloom, A. J., Field, C. B., and Waring, R. H. (1987). Plant responses to multiple environmental factors. *BioScience* **37**, 49–57.

Chew, F. S., and Rodman, J. E. (1979). Plant resources for chemical defense. *In* "Herbivores: Their Interaction with Secondary Plant Metabolites" (G. A. Rosenthal and D. H. Janzen, eds.), pp. 271–307. Academic Press, New York.

Chiariello, N. R., and Gulmon, S. L. (1991). Stress effects on plant reproduction. *In* "Response of Plants to Multiple Stresses" (H. A. Mooney, W. E. Winner, and E. J. Pell, eds.), pp. 161–188. Academic Press, San Diego.

Chiariello, N., and Roughgarden, J. (1984). Storage allocation in seasonal races of an annual plant: Optimal versus actual allocation. *Ecology* 65, 1290–1301.

Clauss, M. J., and Aarssen, L. W. (1994). Phenotypic plasticity of size–fecundity relationships in *Arabidopsis thaliana. J. Ecol.* 82, 447–455.

Cohen, D. (1976). The optimal timing of reproduction. *Am. Nat.* 110, 801–807.

Cohen, Y., and Pastor, J. (1996). Interactions among nitorgen, carbon, plant shape and photosynthesis. Am. Nat. 147, 847–865.

Coleman, J. S., McConnaughay, K. D. M., and Bazzaz, F. A. (1993). Elevated CO_2 and plant nitrogen-use: Is reduced tissue nitrogen concentration size-dependent? *Oecologia* 93, 195–200.

Coley, P. D. (1983). Herbivory and defensive characteristics of tree species in a lowland tropical forest. *Ecol. Monogr.* 53, 209–233.

Coley, P. D. (1986). Costs and benefits of defense by tannins in a neotropical tree. *Oecologia* 70, 238–241.

Coley, P. D., Bryant, J. P., and Chapin III, F. S. (1985). Resource availability and plant antiherbivore defense. *Science* 230, 895–899.

de Jong, T. J. (1986). Effects of reproductive and vegetative sink activity on leaf conductance and water potential in *Prunus persica* cultivar Fantasia. *Sci. Hortic.* (Amsterdam) 29, 131–138.

de Jong, T. J., Klinkhammer, P. G., and Prins, A. H. (1986). Flowering behavior of the monocarpic perennial *Cynoglossum officinale* L. *New Phytol.* 103, 219–229.

de Kroon, H., Fransen, B., van Rheenen, J. W. A., van Dijk, A., and Kreulen, R. (1996). High levels of inter-ramet water translocation in two rhizomatous *Carex* species, as quantified by deuterium labelling. *Oecologia* 106, 73–84.

Detling, J. R., Winn, D. T., Procter-Gregg, C., and Painter, E. L. (1980). Effects of simulated grazing by belowground herbivores on growth, CO_2 exchange, and carbon allocation patterns of *Bouteloua gracilis. J. Appl. Ecol.* 17, 771–778.

Dirzo, R. (1984). Herbivory: A phytocentric overview. *In* "Perspectives in Plant Population Ecology" (R. Dirzo and J. Sarukhán, eds.), pp. 141–165. Sinauer, Sunderland, Massachusetts.

Elmqvist, T., and Gardfjell, H. (1988). Differences in response to defoliation between male and females of *Silene dioica. Oecologia* 77, 225–230.

Fajer, E. D., Bowers, M. D., and Bazzaz, F. A. (1992). The effect of nutrients and enriched CO_2 environments on production of carbon-based allelochemicals in *Plantago:* A test of the carbon/nutrient balance hypothesis. *Am. Nat.* 140, 707–723.

Farnsworth, E. J., and Bazzaz, F. A. (1995). Inter- and intra-generic differences in growth, reproduction, and fitness of nine herbaceous annuals grown in elevated CO_2 environments. *Oecologia* 104, 454–466.

Feeney, P. P. (1976). Plant apparency and chemical defense. *In* "Recent Advances in Phytochemistry, Volume 10: Biochemical Interactions between Plants and Insects" (J. W. Wallace and R. L. Mansell, eds.), pp. 1–40. Plenum, New York.

Field, C. B. (1983). Allocating leaf nitrogen for the maximization of carbon gain: Leaf age as a control on the allocation program. *Oecologia* 56, 341–347.

Field, C. B. (1991). Ecological scaling of carbon gain to stress and resource availability. *In* "Response of Plants to Multiple Stresses" (H. A. Mooney, W. E. Winner, and E. J. Pell, eds.), pp. 35–65. Academic Press, San Diego.

Field, C. B., and Mooney, H. A. (1986). The photosynthesis–nitrogen relationship in wild plants. *In* "On the Economy of Plant Form and Function" (T. J. Givnish, ed.), pp. 25–55. Cambridge Univ. Press, Cambridge.

Fitchner, K., Koch, G. W., and Mooney, H. A. (1995). Photosynthesis, storage and allocation. *In* "Ecophysiology of Photosynthesis" (E.-D. Schulze and M. M. Caldwell, eds.), pp. 133–144. Springer-Verlag, Berlin, Heidelberg, and New York.

Fritz, R. S., and Simms, E. L. (eds.) (1992). "Plant Resistance to Herbivores and Pathogens: Ecology, Evolution, and Genetics." Univ. Chicago Press, Chicago.

Gedroc, J. J., McConnaughay, K. D. M., and Coleman, J. S. (1996). Plasticity in root/shoot partitioning: Optimal, ontogenetic, or both? *Func. Ecol.* **10**, 44–50.

Geiger, D. R., and Servaites, J. C. (1991). Carbon allocation and response to stress. *In* "Response of Plants to Multiple Stresses" (H. A. Mooney, W. E. Winner, and E. J. Pell, eds.), pp. 104–127. Academic Press, San Diego.

Gershenzon, J. (1994). Metabolic costs of terpenoid accumulation in higher plants. *J. Chem. Ecol.* **20**, 1281–1328.

Grime, J. P., and Campbell, B. D. (1991). Growth rate, habitat productivity, and plant strategy as predictors of stress response. *In* "Response of Plants to Multiple Stresses" (H. A. Mooney, W. E. Winner, and E. J. Pell, eds.), pp. 143–159. Academic Press, San Diego.

Gulmon, S. L., and Mooney, H. A. (1986). Costs of defense and their effects on plant productivity. *In* "On the Economy of Plant Form and Function" (T. J. Givnish, ed.), pp. 681–698. Cambridge Univ. Press, Cambridge and New York.

Harper, J. L. (1977). "The Population Biology of Plants." Academic Press, London and New York.

Harper, J. L. (1985). Modules, branches, and the capture of resources. *In* "Population Biology and Evolution of Clonal Organisms" (J. B. Jackson, L. W. Buss, and R. E. Cook, eds.), pp. 1–33. Yale Univ. Press, New Haven, Connecticut.

Hartnett, D. C. (1990). Size-dependent allocation to sexual and vegetative reproduction in four clonal composites. *Oecologia* **84**, 254–259.

Hartnett, D. C., and Bazzaz, F. A. (1985). The integration of neighborhood effects by clonal genets of *Solidago canadensis. J. Ecology.* **73**, 415–427.

Hilbert, D. W. (1990). Optimization of plant root:shoot ratios and internal nitrogen concentration. *Ann. Bot.* **66**, 91–99.

Hirose, T. (1986). Nitrogen uptake and plant growth. II. An empirical model of vegetative and partitioning. *Ann. Bot.* **58**, 487–496.

Hirose, T. (1987). A vegetative plant growth model: Adaptive significance of phenotypic plasticity in matter partitioning. *Funct. Ecol.* **1**, 195–202.

Hirose, T. (1988). Modeling the relative growth rate as a function of plant nitrogen concentration. *Physiol. Plant.* **72**, 185–189.

Hirose, T., and Werger, M. J. A. (1987a). Maximizing daily carbon photosynthesis with respect to the leaf nitrogen allocation pattern in the canopy. *Oecologia* **72**, 520–526.

Hirose, T., and Werger, M. J. A. (1987b). Nitrogen use efficiency in instantaneous and daily photosynthesis of leaves in the canopy of a *Solidago altissima* stand. *Physiol. Plant.* **70**, 215–222.

Hirose, T., Ackerly, D. D., and Bazzaz, F. A. (1996). CO_2 elevation and canopy development in stands of herbaceous plants. *In* "Carbon Dioxide, Populations, and Communities" (Ch. Körner and F. A. Bazzaz, eds.), pp. 413–428. Academic Press, San Diego.

Holland, J. N., Cheng, W., and Crossley, D. A., Jr. (1996). Herbivore-induced changes in plant carbon allocation: Assessment of below-ground fluxes using carbon-14. *Oecologia* **107**, 87–94.

Horwitz, C. C., and Schemske, D. W. (1988). Demographic cost of reproduction in a neotropical herb: An experimental field study. *Ecology* **69**, 1741–1745.

Houghton, J. T., Meira Filho, L. G., Callander, B. A., Harris, N., Kattenberg, A., and Maskell, K. (1995). "Climate Change 1995: The Science of Climate Change. Contribution of Working Group I to the Second Assessment Report of the Intergovernmental Panel on Climate Change." Cambridge Univ. Press, Cambridge, England.

Hutchings, M. J., and de Kroon, H. (1994). Foraging in plants: The role of morphological plasticity in resource acquisition. *Adv. Ecol. Res.* **25**, 159–238.

Iwasa, Y., and Roughgarden, J. D. (1984). Shoot:root balance of plants: Optimal growth of a system with many vegetative organs. *Theor. Popul. Biol.* **25**, 78–105.

Janzen, D. H. (1984). A host plant is more than its chemistry. *Illinois Natural History Survey Bulletin* **33**, 141–174.

Jurik, T. W. (1985). Differential costs of sexual and vegetative reproduction in wild strawberry populations. *Oecologia* **66**, 394–403.

Karban, R., and Adler, F. R. (1996). Induced resistance to herbivores and the information content of early season attack. *Oecologia* **107**, 379–385.

Killingbeck, K. T. (1996). Nutrients in senesced leaves: Keys to the search for potential resorption and resorption proficiency. *Ecology* **77**, 1716-1727.

King, D., and Roughgarden, J. (1982). Graded allocation between vegetative and reproductive growth for annual plants in growing season of random length. *Theor. Popul. Biol.* **22**, 1–16.

Koch, G. W., Schulze, E. D., Percival, F., Mooney, H. A., and Chu, C. (1988). The nitrogen balance of *Raphanus sativus* × *raphanistrum* plants. II. Growth, nitrogen redistribution and photosynthesis under NO_3^- deprivation. *Plant Cell Environ.* **11**, 755–767.

Körner, Ch., and Bazzaz, F. A. (eds.) (1996). "Carbon Dioxide, Population and Communities." Academic Press, London.

Körner, Ch., Scheel, J. A., and H. B. (1979). Maximum leaf diffusive conductance in vascular plants. *Photosynthetica* **13**, 45–82.

Krischik, V. A., and Denno, R. F. (1983). Individual, population and geographic patterns in plant defense. *In* "Variable Plants and Herbivores in Natural and Managed Systems" (R. F. Denno and M. S. McClure, eds.), pp. 463–512. Academic Press, New York.

Labandeira and Phillips (1996). Proc. Nat. Acad. Sci. *U.S.A.*

Lambers, H., and Rychter, A. (1989). The biochemical background of variation in respiration rate: Respiratory pathways and chemical composition. *In* "Causes and Consequences of Variation in Growth Rate and Productivity of Higher Plants" (H. Lambers, M. L. Cambridge, H. Konings, and T. L. Pons, eds.), pp. 199–225. SPB Academic Publishing, The Hague.

Lee, T. D. (1988). Patterns of fruit and seed production. *In* "Plant Reproductive Ecology— Patterns and Strategies" (J. Lovett Doust and L. Lovett Doust, eds.), pp. 179–202. Oxford Univ. Press, New York.

Lee, T. D., and Bazzaz, F. A. (1982a). Regulation of fruit and seed production in an annual legume, *Cassia fasciculata. Ecology* **63**, 1363–1373.

Lee, T. D., and Bazzaz, F. A. (1982b). Regulation of fruit maturation pattern in an annual legume, *Cassia fasciculata. Ecology* **63**, 1374–1388.

Lerdau, M. (1992). Future discounts and resource allocation in plants. *Funct. Ecol.* **6**, 371–375.

Lloyd, D. G. (1980). Sexual strategies in plants I. An hypothesis on serial adjustment of maternal investment during one reproductive session. *New Phytol.* **86**, 69–79.

Loehle, C., and LeBlanc, D. (1996). Model-based assessments of climate change effects of forests: A critical review. *Ecol. Model.* **90**, 1–31.

McConnaughay, K. D. M., and Bazzaz, F. A. (1992). The occupation and fragmentation of space: Consequences of neighbouring shoots. *Funct. Ecology,* **6**, 711–718.

McKey, D. B. (1979). The distribution of secondary compounds within plants. *In* "Herbivores: Their Interaction with Secondary Plant Metabolites" (G. A. Rosenthal and D. H. Janzen, eds.), pp. 55–133. Academic Press, New York.

McNaughton, S. J. (1983a). Compensatory plant growth as a response to herbivory. *Oikos* **40**, 329–336.

McNaughton, S. J. (1983b). Serengeti grassland ecology: The role of composite environmental factors and contingency in community organization. *Ecol. Monogr.* **53**, 291–320.

Marshall, C., and Ellstrand, N. C. (1988). Effective mate choice in wild radish: Evidence for selective seed abortion and its mechanism. *Am. Nat.* **131**, 739–756.

Mattson, W. T. (1980). Herbivory in relation to plant nitrogen content. *Annu. Rev. Ecol. Syst.* **11**, 119–161.

Méndez, M., and Obeso, J. R. (1993). Size-dependent reproductive and vegetative allocation in *Arum italicum* (Araceae). *Can. J. Bot.* **71**, 309–314.

Miao, S. L., Bazzaz, F. A., and Primack, R. (1991). Effects of maternal nutrient pulse on reproduction of two colonizing *Plantago* species. *Ecology* **72**, 586–596.

Michaels, H. J., and Bazzaz, F. A. (1986). Resource allocation and demography of sexual and apomictic *Antennaria parlinii. Ecology* **67**, 27–36.

Michaels, H. J., and Bazzaz, F. A. (1989). Individual and population response breadths of sexual and apomictic plants to environmental gradients. *Am. Nat.* **134**, 190–207.

Mooney, H. A. (1972). The carbon balance of plants. *Annu. Rev. Ecol. Syst.* **XX**, 315–346.

Mooney, H. A., and Gulmon, S. (1979). Environmental and evolutionary constraints on the photosynthetic characteristics of higher plants. *In* "Topics in Plant Population Biology" (O. T. Solbrig, G. B. Johnson, and P. H. Raven, eds.), pp. 316–337. Columbia Univ. Press, New York.

Mooney, H. A., and Gulmon, S. (1982). Constraints on leaf structure and function in reference to herbivory. *BioScience* **32**, 198–206.

Mooney, H. A., and Winner, W. E. (1991). Partitioning response of plants to stress. *In* "Response of Plants to Multiple Stresses" (H. A. Mooney, W. E. Winner, and E. J. Pell, eds.), pp. 129–142. Academic Press, San Diego.

Mooney, H. A., Küppers, M., Koch, G. W., Gorham, J., Chu, C. C., and Winner, W. E. (1988). Compensating effects to growth of carbon partitioning changes in response to SO_2-induced photosynthetic reduction in radish. *Oecologia* **72**, 502–506.

Morse, S. R., and Bazzaz, F. A. (1994). Elevated CO_2 and temperature alter recruitment and size hierarchies in C_3 and C_4 annuals. *Ecology*, **75**, 966–975.

Nitao, J. K., and Zangerl, A. R. (1986). Floral development and chemical defense allocation in wild parsnip (*Pastinaca sativa*). *Ecology* **68**, 521–529.

Nobel, P. S. (1977). Water relations of flowering of *Agave deserti*. *Bot. Gaz.* **138**, 1–6.

Pacala, S., Canham, C., and Silander, J. (1993). Forest models defined by field measurements. 1: The design of a northeastern forest simulator. *Can. J. For. Res.* **23**, 1980–1988.

Parrish, J. A. D., and Bazzaz, F. A. (1985a). Nutrient content of *Abutilon theophrasti* seeds and the competitive ability of the resulting plants. *Oecologia* **65**, 247–251.

Parrish, J. A. D., and Bazzaz, F. A. (1985b). Ontogenetic niche shifts in old-field annuals. *Ecology* **66**, 1296–1302.

Penning de Vries, F. W. T., Brunsting, A. H. M., and van Laar, H. H. (1974). Products, requirements, and efficiency of biosynthesis: A quantitative approach. *J. Theor. Biol.* **45**, 339–377.

Pitelka, L., and Ashmun, J. (1985). Physiology and integration of ramets in clonal plants. *In* "Population Biology and Evolution of Clonal Organisms" (J. B. C. Jackson, L. W. Buss, and R. E. Cook, eds.), pp. 399–436. Yale Univ. Press, New Haven, Connecticut.

Poorter, H. (1994). Construction costs and payback time of biomass: A whole plant perspective. *In* "A Whole Plant Perspective on Carbon–Nitrogen Interactions" (J. Roy and E. Garnier, eds.), pp. 111–127. SPB Academic Publ., The Hague.

Poorter, H., and Bergkotte, M. (1992). Chemical composition of 24 wild species differing in relative growth rate. *Plant Cell Environ.* **15**, 221–229.

Reekie, E. G., and Bazzaz, F. A. (1987a). Reproductive effort in plants. I. Carbon allocation to reproduction. *Am. Nat.* **129**, 876–896.

Reekie, E. G., and Bazzaz, F. A. (1987b). Reproductive effort in plants. II. Does carbon reflect the allocation of other resources? *Am. Nat.* **129**, 897–906.

Reekie, E. G., and Bazzaz, F. A. (1987c). Reproductive effort in plants. III. Effect of reproduction on vegetative activity. *Am. Nat.* **129**, 907–919.

Reekie, E. G., and Bazzaz, F. A. (1992). Cost of reproduction as reduced growth in genotypes of two congeneric species with contrasting life histories. *Oecologia* **90**, 21–26.

Reich, P. B., Walter, M. B., and Ellsworth, D. S. (1992). Leaf life-span in relation to leaf, plant and stand characteristics among diverse ecosystems. *Ecol. Monogr.* **62**, 365–392.

Reinartz, J. A. (1984). Life-history variation of common mullein (*Verbascum thapsus*) I. Latitudinal differences in population dynamics and timing of reproduction. *J. Ecol.* **72**, 897–912.

Reznick, D. (1985). Costs of reproduction, an evaluation of the empirical evidence. *Oikos* **44**, 257–267.

Rhoades, D. F. (1979). Evolution of plant chemical defenses against herbivory. *In* "Herbivores: Their Interaction with Secondary Plant Metabolites" (G. A. Rosenthal and D. H. Janzen, eds.), pp. 3–54. Academic Press, New York.

Robinson, D., and Rorison, I. (1988). Plasticity in grass species in relation to nitrogen supply. *Funct. Ecol.* **2**, 249–257.

Rogers, H. H., Runion, G. B., and Krupa, S. V. (1994). Plant responses to atmospheric CO_2 enrichment with emphasis on roots and the rhizosphere. *Environ. Pollut.* **83**, 155–189.

Rosenthal, G. A., and Janzen, D. H. (eds.) (1979). "Herbivores: Their Interaction with Secondary Plant Metabolites." Academic Press, New York.

Rupp, M. J., and Denno, R. F. (1983). Leaf age as a predictor of herbivore distribution and abundance. *In* "Variable Plants and Herbivores in Natural and Managed Systems" (R. F. Denno and M. S. McClure, eds.), pp. 91–124. Academic Press, New York.

Sage, R. F. (1994). Acclimation of photosynthesis to increasing atmospheric CO_2: The gas exchange perspective. *Photosynth. Res.* **39**, 351–368.

Schmid, B. (1990). Some ecological and evolutionary consequences of modular organization and clonal growth in plants. *Evolutionary Trends in Plants* **4**, 25–34.

Schmid, B., and Weiner, J. (1993). Plastic relationships between reproductive and vegetative mass in *Solidago altissima. Evolution* **47**, 61–74.

Schmid, B., Bazzaz, F. A., and Weiner, J. (1995). Size dependence of sexual reproduction and clonal growth in two perennial plants. *Can. J. Bot.* **73**, 1831–1837.

Schulze, E.-D. (1991). Water and nutrient interactions with plant water stress. *In* "Response of Plants to Multiple Stresses" (H. A. Mooney, W. E. Winner, and E. J. Pell, eds.), pp. 89–101. Academic Press, San Diego.

Schulze, E., and Chapin, F. S. I. (1987). Plant specialization to environments of different resource availability. *In* "Potentials and Limitations of Ecosystem Analysis" (E.-D. Schulze and H. Zwolfer, eds.), pp. 120–148. Springer-Verlag, Berlin.

Schulze, W., and Schulze, E.-D. (1995). The significance of assimilatory starch for growth in *Arabidopsis thaliana* wild-type and starchless mutants. *In* "Ecophysiology of Photosynthesis" (E.-D. Schulze and M. M. Caldwell, eds.), pp. 123–131. Springer-Verlag, Berlin, Heidelberg, and New York.

Solbrig, O. T. (1981). Studies on the population biology of the genus *Viola.* II. The effect of plant size on fitness in *Viola sororia. Evolution* **35**, 1080–1093.

Stephenson, A. G. (1981). Flower and fruit abortion: Proximate causes and ultimate functions. *Annu. Rev. Ecol. Syst.* **12**, 253–279.

Stulen, I., and der Hertog, J. (1993). Root growth and functioning under atmospheric CO_2 enrichment. *Vegetatio* **104/105**, 99–115.

Sultan, S. E. (1996). Phenotypic plasticity for offspring traits in *Polygonum persicaria. Ecology* **77**, 1971–1807.

Thornley, J. H. M. (1991). A model of leaf tissue growth, acclimation and senescence. *Ann. Bot.* **67**, 219–228.

Traw, M. B., Lindroth, R. L., and Bazzaz, F. A. (1996). Decline in gypsy moth (*Lymantria dispar*) performance in an elevated CO_2 atmosphere depends upon host plant species. *Oecologia* (in press).

Tremmel, D. C., and Bazzaz, F. A. (1993). How neighbor canopy architecture affects target plant performance. *Ecology* **74**, 2114–2124.

Tremmel, D. C., and Bazzaz, F. A. (1995). Plant architecture and allocation in different neighborhoods: Implications for competitive success. *Ecology* **76**, 262–271.

van den Driessche, R., and Ponsford, D. (1995). Nitrogen induced potassium deficiency in white spruce (*Picea glauca*) and Engelmann spruce (*Picea engelmannii*) seedlings. *Can. J. For. Res.* **25**, 1445–1454.

Vitousek, P. M. (1994). Beyond global warming: Ecology and global change. *Ecology* **75**, 1861–1876.

Waring, R. H., and Schlesinger, W. H. (1985). "Forest Ecosystems: Concepts and Management." Academic Press, Orlando, Florida.

Wayne, P. M., and Bazzaz, F. A. (1993a). Morning vs. afternoon sun patches in experimental forest gaps: Consequences of temporal incongruency of resources to birch regeneration. *Oecologia* **94**, 235–243.

Wayne, P.M., and Bazzaz, F. A. (1993b). Birch seedling responses to daily time courses of light in experimental forest gaps and shadehouses. *Ecology* **74**, 1500–1515.

Werner, P. A. (1979). Competition and coexistence of similar species. *In* "Topics in Plant Population Biology" (S. K. Jain, O. T. Solbrig, G. B. Johnson, and P. H. Raven, eds.), pp. 287–312. Columbia Univ. Press, New York.

Whittaker, R. H. (1966). Forest dimensions and production in the Great Smoky mountains. *Ecology* **47**, 103–121.

Williams, K., Percival, F., Merino, J., and Mooney, H. (1987). Estimation of tissue construction cost from heat of combustion and organic nitrogen content. *Plant Cell Environ.* **10**, 725–734.

Willson, M. F. (1983). "Plant Reproductive Ecology." Wiley, New York.

Wilson, J. B. (1988). A review of evidence on the control of shoot/root ratio in relation to models. *Ann. Bot.* **61**, 433–449.

Zangerl, A. R., and Bazzaz, F. A. (1992). Theory and pattern in plant defense allocation. *In* "Plant Resistance to Herbivores and Pathogens" (R. S. Fritz and E. L. Simms, eds.), pp. 363–391. Univ. of Chicago Press, Chicago.

2

The Fate of Acquired Carbon in Plants: Chemical Composition and Construction Costs

Hendrik Poorter and Rafael Villar

I. Introduction

One of the goals of ecophysiology and agronomy is to understand the physiological basis of plant growth. In this respect, the process of photosynthesis has received ample attention. Large efforts have been made to analyze the physical and biochemical processes that are necessary for carbon fixation. The subsequent fate of the acquired carbon (C) has been investigated far less. Factors controlling the plant's respiration and translocation of C to several organs are only fragmentarily understood. This applies even more so to the fate of C in a specific organ or cell. In most studies related to plant growth, biomass accumulation is taken as such and not analyzed further. However, the way a plant invests its carbon as well as the other acquired elements into different compounds may have a profound effect on its growth and performance in a certain environment (e.g., Bazzaz *et al.*, 1987). Therefore, it is crucial to understand the factors controlling chemical composition.

So far, we only have a limited understanding of the causes and consequences of variation in chemical composition. In this chapter, we discuss first the integration level at which the chemical composition of plants will be considered. Second, we characterize the chemical composition of various plant organs at that integration level. What values for the concentrations of the various plant compounds could be taken as typical for an "average" plant, and what variation is to be expected around these average values? Third, we analyze to what extent these compounds covary in specific patterns. That is, in comparing different species or a range of environments, will a rise in compound A always be accompanied by a decrease in compound B? Or do both compounds vary independently? Fourth, we review the possible mechanistic explanations that have been put forward to explain variation in chemical composition.

The last two sections of this chapter discuss the consequences of variation in chemical composition. Given the chemical composition of a plant, it is possible to arrive at an estimate of the total amount of photosynthate that has to be spent to construct one gram of biomass: the so-called construction costs. First, we focus on these construction costs, and discuss to what extent they depend on environment and type of species. Second, we briefly discuss the ecological consequences of variation in chemical composition.

II. Integration Level

Plants contain a vast range of compounds, with estimates of more than 100,000 present (Buckingham, 1993), most of them in very small amounts. A complete analysis of all these compounds is impossible. Although powerful techniques are available now to analyze small samples on large ranges of constituents (e.g., pyrolysis–mass spectrometry; Boon, 1989), such a detailed picture would not be of much help to understand processes at a higher integration level. Rather, it is preferable to categorize these compounds in a limited number of classes of constituents, which yields the "proximate" chemical composition (cf. Penning de Vries et al., 1974). In this chapter we use eight different categories: (1) lipids, (2) lignin, (3) soluble phenolics (tannins, flavonoids), (4) organic N compounds (which we will call "protein" throughout this chapter but which consist of at least DNA, RNA, chlorophyll, and amino acids as well), (5) total structural carbohydrates (TSC: cellulose, hemicellulose, and pectin), (6) total nonstructural carbohydrates (TNC: starch, fructan, sucrose, fructose, glucose), (7) organic acids, and (8) minerals. In general, these compounds together comprise more than 90–95% of a plant's biomass (Chapin, 1989; Poorter and Bergkotte, 1992). The other constituents of a plant are mostly present in only small concentrations. However, in some species compounds like cyanogenic glucosides (Merino et al., 1984) or terpenes (Bryant et al., 1983) are found in relatively large amounts (>5%) of the plant's biomass. This classification is very broad and so may not be suitable for some kinds of ecophysiological problems. For example, in cases where water relations are of interest, it may be useful to separate the soluble fraction from the insoluble part of the TNC, and combine the soluble sugars with the organic acids, as both have a similar osmotic function.

Generally, not all of the eight groups of compounds are determined on the same biological samples. Although this is not always necessary to answer a specific question, there is an added value in knowing the overall proximate chemical composition of the various organs of a plant. In this way, trade-offs between allocation of C and nutrients to constituents with different functions in the plant can be evaluated.

Concentrations of chemical compounds in plants can be expressed in various ways. In the literature, values on either a fresh or dry weight basis are used, per unit of leaf area or as a ratio, relative to another compound. Moreover, concentrations are analyzed for integration levels varying from organelles up to that of the whole plant. In this chapter we follow the majority of papers and express concentrations per unit dry weight. This avoids the problem of variation in dry weight:fresh weight ratios or leaf area:leaf weight ratios that could mask genuine differences in concentrations. However, depending on the context of the research, other expressions may be more useful in specific cases. Concerning the integration level within the plant we differentiate between leaves, stems, roots, as far as vegetative organs are concerned, and seeds and fruit flesh as the two major components of reproductive organs. We believe that concentrations expressed on a whole plant weight basis may be helpful as well in understanding the growth and functioning of plants (see Poorter, 1994). However, due to differences in composition between organs, concentrations per unit total plant are confounded with the allocation pattern of biomass to the different organs, and we will not use it here.

In Appendix 1, a procedure for a proximate analysis is presented, as used by one of us (Poorter and Bergkotte, 1992). Reference is made to the various determinations necessary. Appendix 2 lists the subsequent assumptions and calculations to arrive at an estimate of the different classes of compounds. Another scheme for a more or less complete extraction is given by Kedrowski (1983).

III. Chemical Composition

Before being able to discriminate between "low" and "high" values for the different constituents, we have to establish "normal" concentrations and their variation. Only a few attempts have been made to fully characterize the proximate chemical composition of a plant. These do not suffice to obtain a general picture. Therefore, we screened the literature for determinations of each of the eight groups, compiling data from a wide variety of sources. Median values, as well as the ranges generally found, are given in

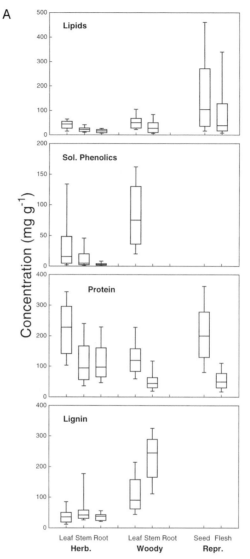

Figure 1   Characterization by box plots of the chemical composition of various plant tissues for (A) relatively expensive classes of compounds and (B) relatively inexpensive classes of compounds. All concentrations are expressed in milligrams per gram dry weight. Data are categorized into values pertaining to herbaceous species (leaf, stem, root), woody species (leaf, stem, root), and categories of seeds and fruit flesh. Values are extracted from a wide range of literature sources (among others, Bliss, 1962; Loveless, 1962; Caspers, 1977; Duke and Atchley, 1986; Chapin, 1989; Fengel and Wegener, 1989; Poorter and Bergkotte, 1992; Jordano, 1995). No data are presented in categories in which we collected less than 20 observations. The number of observations on concentrations in leaf, stem, and roots

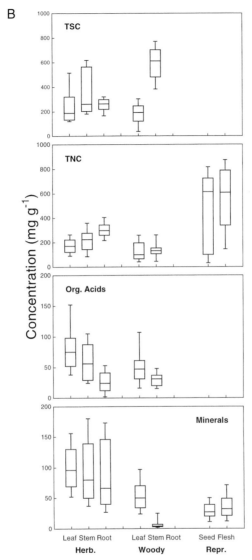

of herbaceous plants and leaves, stems, and roots of woody species, as well as for seeds and fruit flesh, are as follows: lipids, 190,50,80/200,60,—/320,490; lignin, 80,40,30/100, 180,—/—,—; soluble phenolics, 70,40,30/110,—,—/—,—; protein, 360,150,160/260,90,—/ 340,560; TSC, 100,50,30/80,90,—/—,—; TNC, 190,50,100/80,20,—/60,460; organic acids, 130,30,30/30,20,—/—,—; minerals: 420,50,90/250,160,—/320,260. Box plots indicate the distribution by percentiles. The *x*th percentile is the value below which *x*% of the observations are found. The lower and higher part of the box indicate the 25th and 75th percentiles, respectively. The value of the error bars are the 10th and 90th percentile, and the 50th percentile (median) is given by the horizontal line within the box.

Fig. 1. It should be stressed that the compiled values were obtained with different methods of quantification, for plants grown under a wide range of conditions. Especially the concentrations of protein, TNC, and minerals will depend strongly on the levels of resource supply (Waring *et al.*, 1985; McDonald *et al.*, 1986; Griffin *et al.*, 1993). Nevertheless, Fig. 1 provides a useful indication of the concentrations that are to be expected. In leaves of herbaceous species, median values are highest for protein followed by TSC, TNC, and minerals, each of which comprise 10% or more of the leaves' biomass. Organic acids, lipids, and lignin show median concentrations around 5%, and levels of soluble phenolics are just a few percent. Stems and roots of herbaceous plants have lower concentrations of protein, organic acids, soluble phenolics, and lipids, and higher values for TSC and TNC than leaves. Compared to leaves of herbaceous species, leaves of woody species contain less proteins, minerals, and organic acids, similar amounts of lipids, TNC, and TSC, and higher concentrations of lignin and soluble phenolics. As far as information is available, the same pattern is found when stems from herbaceous species are compared with those of woody species, with the notable exception of TSC, which shows much higher concentrations in stems of woody plants. We were not able to find enough data for a reliable estimate of the chemical compositions of roots of woody species. Judging from available information, they are quite similar to those of woody stems. On average, seeds and fruit flesh are characterized by rather high concentrations of TNC and lipids, and lower concentrations of minerals. A difference between the two is that seeds can have high concentrations of protein, whereas this is not the case in fruit flesh. Variation in chemical composition of the reproductive plant parts is much larger than in vegetative organs. This obviously reflects the function of these plant parts in accumulating storage compounds of various types, or attracting different animal seed dispersers (see Jordano, 1995).

In the above compilation, data from a wide range of species and growth conditions were combined. To what extent do specific investigations on differences between organs or species support the above trends? Poorter and Bergkotte (1992) analyzed the levels of the eight classes of compounds described above for leaves, stems, and roots of 24 herbaceous species. For each species, the concentration in stems, and roots was calculated relative to that in the leaves. Subsequently, for an overall impression these values were averaged over all species investigated (Table I). In both stems and roots, concentrations of lipids, soluble phenolics, protein, and organic acids were lower, whereas those of lignin, TSC, TNC, and minerals were higher than in leaves. As far as determined, these observations are in line with those of Niemann *et al.* (1993) on tomato and of Challa (1976) on cucumber. We are not aware of data on whole stems and roots of woody species. An interesting aspect shown in the data of Table I is that differences in chemical

Table I Ratios of Concentrations of Compounds in Stems and Roots, Relative to Those of Leaves[a]

Compound	Stem	Roots
Lipids	0.57***	0.49***
Organic acids	0.77**	0.28***
Soluble phenolics	0.77*	0.64**
Protein	0.80***	0.70***
Minerals	1.27***	1.46***
TSC	1.39***	1.69***
Lignin	1.78***	1.69***
TNC	1.85***	2.15***

[a] Average values of data on 24 herbaceous species (Poorter and Bergkotte, 1992). Values are back-transformed averages after natural log transformation of the ratios, to correct for the ln-normal distribution of ratios. Asterisks indicate to what extent the average deviates from 1.0 (H_0 hypothesis, no difference between leaves and stems or roots): *, $P < 0.05$; **, $P < 0.01$; ***, $P < 0.001$.

composition between stems and roots are small, when they are compared with those of leaves. Clearly, if one prefers to analyze plant composition in a two- rather than three-compartment model, it is advisable to separate leaves from stems and roots, rather than separating aboveground and belowground plant parts.

Although there is common knowledge about how groups of species differ in composition, we are not aware of many larger scale comparisons to support these notions with a quantitative basis. Poorter and Bergkotte (1992) and Niemann et al. (1992) found higher concentrations of protein, organic acids, and minerals in potentially fast-growing species, and higher concentrations of cell-wall compounds in herbaceous species with a low growth potential (see also Section IV). In a comparison of four woody and six herbaceous species, grown at a nonlimiting nutrient supply, a clear difference was found in the concentration of minerals (H. Poorter and J. R. Evans, unpublished). In particular, NO_3^- was present in small concentrations in the leaves only of woody plants. Differences in organic acids and proteins showed similar directions as those in Fig. 1, although the differences were rather small. The more clear-cut differences in Fig. 1 may be a reflection of the fact that most data on woody species are from material collected in the field. Within the group of the woody plants, evergreen and deciduous species differ in their chemical composition. Leaves of deciduous plants have higher concentrations of proteins (Loveless, 1962) and minerals than leaves of evergreens (R. Villar and J. Merino, unpublished). The concentration of lipids were found to be similar.

Environmental impact on chemical composition is especially large for protein, lignin, minerals, and TNC. High-light conditions result in high concentrations of lignin, whereas low-light plants accumulate more minerals (Waring *et al.*, 1985; H. Poorter and J. R. Evans, unpublished). Plants grown at high nutrient levels have higher concentrations of protein, and minerals and lower levels of TNC (Waring *et al.*, 1985; McDonald *et al.*, 1986). Plants at high CO_2 accumulate TNC to a large extent, and apart from this show lower protein and mineral concentrations. Lignin concentrations are hardly affected (Körner *et al.*, 1995; Poorter *et al.*, 1997).

IV. Covariation in Plant Compounds

Given the ranges that can be expected in the various compounds (Fig. 1), the next question to address is to what extent compounds covary. That is, will high values of constituent A always be accompanied by a high (or a low) concentration of constituent B? Insight into these patterns is required before a proper analysis of the regulation of chemical composition can be made. This question about covariation can be answered at two levels, a methodological and a biological one. Concentrations of compounds are generally expressed per unit total dry weight. Therefore, in the end the total composition will always add up to a concentration of 100%. Consequently, it would be expected that different plant compounds covary in a negative way, due to a change in one compound only. For example, an increase in the concentration of compound A from 100 to 250 mg g^{-1} will automatically imply a decrease in all other compounds of 13%. As there is no "internal standard" to which to relate changes in composition, there is no straightforward solution to this problem. In cases where large absolute changes are to be expected, like the accumulation of starch in the leaves of plants grown at elevated CO_2 (e.g., Körner *et al.*, 1995), it may be worth considering concentrations on a TNC-free basis. Alternatively, concentrations have been expressed per unit fresh weight. As the dry biomass generally comprises just a small part of the total fresh weight, changes in one compound will only have a small effect on changes in another. However, with such a solution it is implicitly assumed that there is hardly any difference in the water content per unit dry weight. Any variation in water content between organs or species will confound the results.

Apart from the methodological, negative correlation mentioned above, covariation can have a biological background. For example, if plants are grown at a range of N availabilities, there is generally a negative correlation between the amount of protein and TNC, even if the amount of protein

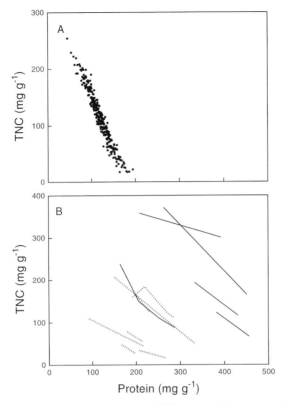

Figure 2   Relationship between the amount of TNC (or starch) and protein (on a TNC-free basis) in leaves or shoots of various plant species. (A) Two thousand observations on *Oryza sativa* (Batten *et al.*, 1993). All values were ranked on the basis of their TNC value. Data points give the average values for 10 consecutive individuals in the ranking. (B) Data on several herbaceous (continuous lines) and woody (broken lines) species, grown at various levels of N availability. Data are from Hehl and Mengel (1972), Waring *et al.* (1985), McDonald *et al.* (1986), Landsberg (1987), Wong *et al.* (1992), and Mooney *et al.* (1995).

is expressed on a TNC-free basis (Fig. 2). This is an intriguing phenomenon, which is not well understood. Given the high concentration of nonstructural carbohydrates in the leaves at low N availability, we can deduce that the amount of photosynthates fixed during photosynthesis is in itself not the factor limiting growth of these plants. Clearly, low-N plants have an excess of sugars. If it is not the amount of photosynthetic machinery that limits the growth at N limitation, it must be the amount of N invested in nonphoto-synthetic compounds. Apparently, at a low nitrogen availability, plants over-invest N in compounds related to the photosynthetic machinery. If we would assume that plants try to maximize their growth in a given set of limiting resources, it has to be concluded that these low-N plants generally behave suboptimally under such conditions. As yet, we have no idea about the reason for this behavior.

Covariation in chemical composition is not only environmentally induced, but could also be due to inherent differences between species. As noted above, even for different species grown under the same conditions, variation in chemical composition is considerable. Poorter and Bergkotte (1992) analyzed the chemical composition of leaves, stems, and roots of 24 wild herbaceous species, differing in potential relative growth rate. How do the different compounds covary in these cases? To investigate this we carried out a factor analysis (Fig. 3). This is a way to characterize how well related a range of variables is (see the legend to Fig. 3 for an explanation). As can be seen, there is a general clustering of cytoplasmic and vacuolar compounds (protein, minerals, organic acids) on one hand (see Poorter, 1994) and cell-wall compounds (lignin, TSC) on the other hand. This is true for all the organs of these species (Fig. 3A,C,D). Positively correlated with the cytoplasmic/vacuolar complex are the total N concentration, the potential growth rate of the investigated species, the leaf area : leaf dry weight ratio (specific leaf area, SLA), as well as the water content per unit dry weight. Positively correlated with the cell-wall complex is the C concentration.

How general are the observed trends? Do they still hold when species of different life forms are compared? Poorter *et al.* (1997) analyzed the proximate chemical composition of leaves of 7 trees, 9 crop species, and 11 wild herbaceous plants. Only three of these were used in the previous experiment as well. All species were grown under controlled conditions with a relatively high nutrient supply, but in different laboratories. Therefore, light conditions, temperature, and soil substrate may have varied to some extent. It is striking how similar the results (Fig. 3B) are when these are compared with the leaves of the 24 herbaceous species (Fig. 3A), with again a clustering of protein, minerals, and organic acids on the one hand, and lignin and soluble phenolics on the other. The exceptions are the lipids and TSC, which have swapped places. The strong correlation of lipids with the "slow-growth/low SLA" complex is due to the evergreen *Eucalyptus* species within the data set, which showed higher concentrations of lipids ($50–100$ mg g^{-1}) than the herbaceous species. Relatively high values of lipids in leaves of evergreen species as compared to deciduous shrubs and trees have also been found in a large survey of field-grown species (R. Villar and J. Merino, unpublished). That survey showed a positive correlation between protein and minerals within the woody species as well. We are not aware of many experiments where a wide range of parameters have been

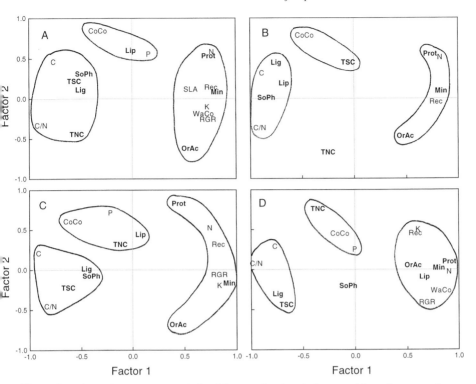

Figure 3 Principal component analysis of data on the chemical composition of a range of plant species. (A) Leaves of 24 different herbaceous plant species, varying in relative growth rate (RGR). RGR, specific leaf area (SLA, leaf area/leaf dry weight), and water content (g water/g dry weight) of the leaves are included as well. (B) Leaves of 27 species (7 woody, 9 crop species, and 11 herbaceous wild plants), all grown at high nutrient availability. (C) Stems and (D) roots of 24 different herbaceous plants, varying in potential relative growth rate. RGR and water content of the stems and roots, respectively, are included as well. The two factors explained 50–60% of the total variation. Data of A, C, and D are from Poorter and Bergkotte (1992), and data of B are from Poorter *et al.* (1997). Abbreviations: CoCo, construction costs; Lig, lignin; Lip, lipids; Min, minerals; OrAc, organic acids; Prot, protein; Rec, recovery (total fraction of the biomass explained by the sum of all concentrations); SoPh, soluble phenolics; TSC, total structural carbohydrates; TNC, total nonstructural carbohydrates; WaCo, water content. Abbreviations in bold type pertain to the eight classes of compounds indicated in Section II. In this analysis two new variables (factor 1 and factor 2) are computed out of a combination of all original variables. For each of these variables it is calculated whether they contribute positively (close to 1.0), negatively (close to −1.0), or not (close to 0.0) to factor 1. The amount of variance thus explained is taken out of the data, and the procedure is repeated with the remaining variance. The result is somewhat comparable to a two-dimensional electrophoresis. Variables that are close together (like total N and protein in Fig. 3B) are generally positively correlated, variables that are at opposite parts of the graph (like minerals and soluble phenolics) are negatively correlated, and variables that have values close to 1 or −1 for one factor and values close to 0 for the other axis (like minerals and, to some extent, TSC) are generally not correlated at all.

determined simultaneously. Dijkstra (1989) found similar correlation patterns as in Fig. 3 for two *Plantago major* subspecies, differing in relative growth rate (RGR), SLA, leaf water content, TNC, minerals, and total cell-wall material. At variance with the trends observed here, protein concentration did not differ for these subspecies.

It is noteworthy that these patterns of investment, either with high concentrations of proteins, minerals, and organic acids or high concentrations of lignin and cell-wall components, coincide with the concentration of total C and N, and with the C/N ratio. By definition, C is positively and N negatively correlated with the C/N ratio (cf. panels in Fig. 3). However, as variation in the C concentration is generally confined to the range 400–500 mg g^{-1}, whereas N may vary more than fivefold, C/N ratios mainly depend on variation in N. Lignin and TSC generally cluster together with C and the C/N ratio, but vary more than C (Fig. 3). This supports the use of parameters like the crude fiber:protein ratio (Loveless, 1962) or lignin:N as applied in decomposition studies (Berg and Staaf, 1980; Melillo *et al.*, 1982). However, if these parameters are not available, C/N ratios are useful descriptors of plant material. An advantage of the use of the C/N ratio is that it can be determined easily with an elemental analyzer, independently of extractions and colorimetric or enzymatic reactions. Thus, interactions of the determination with other compounds, which generally result in lower estimates, do not occur. Such drawbacks are characteristic of the usual proximate analyses (see Poorter, 1994). Note, for example, that the sum of the amount of dry weight ascribed to the various classes of compounds (indicated as "recovery") is correlated positively with RGR and water content, but negatively with investment in cell walls (Fig. 3). This could be due to the fact that slow-growing species do have compounds we did not analyze for (like the terpenes mentioned above). Alternatively, and most likely, the chemical assays may have been disturbed by the wealth of secondary compounds in the plant material.

V. Mechanistic Explanations for Variation in Chemical Composition

What is the reason for the emerging patterns of compounds within a given organ? Which physiological and/or morphological characteristics determine whether a plant will have a low or a high concentration of a given constituent? The underlying mechanisms are not well understood. In this section, we discuss a number of possible explanations. In some cases we are only able to correlate chemical composition with some other plant traits, without understanding the mechanisms behind it. In searching for explanations it is necessary to partly deviate from the distinction of the eight categories we have made, and to consider specific compounds and processes at a more detailed level.

A first explanation for the negative correlation between protein and minerals, on the one hand, and cell walls, on the other, may simply be ontogeny (Chapin, 1989). Young material, just formed, will have high concentrations of protein, whereas older tissue has undergone deposition of secondary cell walls and will show higher concentrations of lignin and TSC. If fast-growing species had relatively large amounts of young tissue, as compared to slower growing plants, this might explain the negative correlation.

A second cause for the distinction between species investing in cytoplasmic/vacuolar compounds and those investing in cell-wall compounds may be a difference in anatomy at the cellular level. If large numbers of sclerenchyma cells are formed, this will obviously increase the amount of lignin and TSC relative to that of protein (see Garnier and Laurent, 1994; Van Arendonk and Poorter, 1994). In addition, relatively high proportions of cell-wall compounds are also expected in small-sized cells with a high cell wall area : cell volume ratio.

To some extent, differences in the concentration of starch between species are correlated with the mode of phloem loading. That is, species that have a symplastic type of phloem loading have much higher starch concentrations in the leaves than species that load apoplastically (Van Bel, 1994; Körner *et al.*, 1995). The cause for the difference in TNC accumulation between the two groups has not yet been clarified.

It has been suggested that lignification could hinder a high metabolic activity, due to a low water permeability of the cell wall, and that this would be a reason for high-protein plants having low concentrations of lignin (Chapin, 1989). We consider this explanation to be unlikely. Lignin is generally accumulated in cells that are dead, or at best have a marginal metabolic activity (sclerenchyma, tracheids, xylem vessels). Cells with a high metabolic activity (palisade and spongy parenchyma, accompanying cells) do not show significant lignification.

Other explanations for the observed patterns are at the biochemical/ physiological level. A clear cause and effect can explain the positive correlation between the concentrations of protein and organic acids in the leaf. According to the Benzioni–Lips model, NO_3^- which is transported to the leaf is reduced there. The negative charge of the nitrate is transferred to an organic acid (malate), which will partly be transported to the roots and broken down, and partly accumulated in the vacuoles of leaf cells (Dijkshoorn *et al.*, 1968). Thus, a rather close correlation between the concentrations of organic acid and protein is expected. Indeed, both load high on the first axis in the factor analysis (Fig. 3A,B). However, it is noteworthy that in both screening experiments there is a separation between protein and organic acids, when the second factor is considered. We have no explanation for this pattern, which is observed in stems as well (Fig. 3C). A close correlation between leaf protein and organic acids is not expected either in plants that predominantly reduce nitrate in the root, such as tree species (Gojon *et al.*, 1994), or in leaves of plants grown with ammonium (Dijkshoorn *et al.*, 1968; Raven, 1985).

Apart from the differences due to inherent variation in leaf anatomy, the relative investment in cytoplasmic/vacuolar compounds versus cell-wall compounds may also be regulated at the biochemical level. Two major theories have been put forward. In the first, synthesis of quantitative secondary compounds is thought to be regulated by the amount of sugars available. In the case of, for example, low nutrient availability, but also for inherently slow-growing species, total nonstructural carbohydrates accumulate (Figs. 2 and 3), which may trigger incorporation of C in carbon-based compounds (Bryant *et al.*, 1983), largely invested in cell walls. Alternatively, and more specifically focused on compounds of a phenolic nature, there may be some kind of competition between synthesis of proteins and that of phenolics. In both pathways the amino acids tyrosine and phenylanaline play a role. At high levels of protein production, the amino acids would be readily incorporated into proteins. If protein synthesis is low, however, phenolics may be produced (Margna 1977). Lambers (1993) reviews both theories and concludes that the experimental evidence up to now is mainly correlative.

Fast-growing species have higher amounts of water per unit dry weight, and therefore require relatively large amounts of osmotics. Consequently, high concentrations of minerals, soluble sugars, and/or organic acids per unit dry weight are found (see Fig. 3). The exact osmotic used may be a regulation point by itself. For example, Blom-Zandstra *et al.* (1988) found that one genotype of *Lactuca sativa* accumulates relatively high concentrations of organic acids in its vacuoles, whereas another mainly uses nitrate. The latter was growing fastest, possibly reflecting the fact that a larger part of the fixed C could be invested in structural growth (see Raven, 1985). Similarly, there could be a regulation mechanism depending on light intensity: the lower the light intensity, the more of the osmotically active sugars are replaced by nitrate (Wedler, 1980).

Candidates for a possible regulatory role are the hormones. However, we are not aware of many studies that have investigated this aspect. Niemann *et al.* (1993) studied the chemical composition of wild-type tomatoes and mutants deficient in gibberellic acid (GA). The mutants had higher concentrations of protein and lower concentrations of cellulose. This might be related to the relatively larger cells of the GA-deficient mutants. These changes may therefore be a rather indirect effect of hormone production.

VI. Construction Costs

A. Carbon Budget

What are the consequences of differences in chemical composition for plant growth? There are two ways to approach this problem. First, one can construct a carbon budget. Such a budget relates quantitatively to the relative growth rate of a plant:

$$RGR = \frac{PS_a \times SLA \times LWR - LR_w \times LWR - SR_w \times SWR - RR_w \times RWR}{C_L \times LWR + C_S \times SWR + C_R \times RWR}, \quad (1)$$

where PS_a is the rate of photosynthesis per unit leaf area and LR_w, SR_w, and RR_w are the rate of respiration per unit leaf, stem, and root weight, respectively. All these values are expressed as fluxes of C integrated over a day. LWR, SWR, and RWR are the fractions of biomass allocated to leaves, stems, and roots, respectively, SLA is the specific leaf area, and C_L, C_S, and C_R are the concentrations of carbon per unit dry weight of the three organs. The total carbon gain of a plant, expressed per unit total plant weight and integrated over a day, is represented by Fig. 4A. A fraction of the C fixed over that day is respired again. This respiration is used to provide energy and/or reducing power for maintenance, to take up minerals by the roots, for transport of compounds, as well as for growth (Van der Werf *et al.*, 1994). The remaining fraction is considered to be C invested in new growth as C skeletons. However, the actual amount of biomass which can be formed

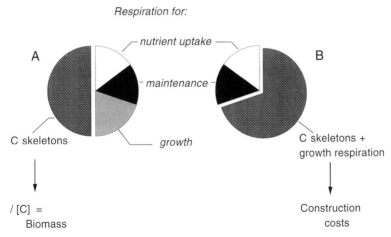

Figure 4 Representation of the total carbon gain and the fate of the fixed glucose. (A) Separation into glucose C spent in respiration and C invested in C skeletons. (B) Separation into glucose C spent in respiration for maintenance and uptake of nutrients, and construction costs (growth respiration plus C skeletons).

will depend on the carbon concentration of that biomass. Generally, the C concentration varies from 400 mg g^{-1} in herbaceous plants grown hydroponically up to 550 mg g^{-1} in highly lignified trees (Poorter, 1989). Consequently, everything else being equal, it can be derived from Eq. (1) that relative growth rate may vary by 38% due to variation in the amount of C that is present in 1 gram of biomass.

A second avenue to analyze the growth of plants is to split up the C-requiring processes in a slightly different way, by adding the C spent in growth respiration to the C invested in C skeletons (Fig. 4B). In this way, we arrive at the construction costs of a plant, which are defined as the amount of glucose required to construct 1 gram of biomass (Penning de Vries *et al.*, 1974; Williams *et al.*, 1987). Consequently, this value not only includes the glucose for providing C skeletons, but also the NAD(P)H and ATP to drive the energy-requiring reactions (Fig. 4).

How do construction costs relate to the chemical composition, as discussed in the previous sections? Taking the most likely biosynthetic pathways, the amount of glucose required to build 1 gram of any of the eight classes of compounds can be calculated (Penning de Vries *et al.*, 1974; Lambers and Rychter, 1989). These values, which include the reducing power and ATP necessary, are listed in Table II. The costs of uptake and transport are accounted for in Fig. 4 by respiration for uptake. Consequently, the construction costs for minerals are nil. Within the group of organic compounds, differences amounting to a factor of 3 are present. Clearly, the construction costs of the plant will depend on the relative contribution of expensive compounds, like lipids, lignin, soluble phenolics, and protein, on the one hand, and the level of cheap compounds, like TSC, TNC, organic acids, and minerals, on the other.

Table II   Amount of Glucose Required for and Amount of CO$_2$ Produced during Synthesis of 1 Gram of Different Compounds[a]

Compound	Construction costs (g glucose g^{-1})	CO$_2$ produced (mmol g^{-1})
Lipids	3.03	36.5
Soluble phenolics	2.60	31.9
Protein (with NO$_3^-$)	2.48	37.9
Lignin	2.12	13.1
TSC	1.22	2.8
TNC	1.09	1.8
Organic acids	0.91	−1.0
Minerals	0	0

[a] Data from Penning de Vries *et al.* (1983) and Lambers and Rychter (1989).

B. Limitations

There are a number of technical as well as conceptual limitations related to the determination and use of construction costs in ecophysiological research. Technical details and assumptions are discussed by Chiariello *et al.* (1989) and Poorter (1994). Here we concentrate on the conceptual issues. First, the costs of the various compounds are based on the most likely biochemical pathways. It is not necessarily true that biosynthesis follows these routes. We are not aware of a systematic evaluation of possible errors involved. Penning de Vries *et al.* (1974) assessed variation in construction costs for different constituents within several of the classes of compounds distinguished here. Differences were generally within the 5% range and were small compared to differences between classes of compounds.

Second, it is assumed that the respiration involved in growth processes has a P/O ratio of 3. That is, when respiration is running most efficiently, three molecules of ATP are formed per oxygen atom (O) consumed. However, if the alternative pathway of respiration is operational, the P/O ratio will decrease. The alternative pathway of respiration yields only one ATP per O consumed (Lambers, 1985). If the alternative pathway respiration represents half of the total respiration rate, the P/O ratio will decrease to about 2. Consequently, more glucose is required to provide for the accessory energy. The exact amount of extra glucose depends on the chemical composition of the plant. For the range of woody, wild herbaceous, and crop species from Fig. 3B, the change in P/O ratio from 3 to 2 would lead to a 4.5–6% increase in construction costs (see Lambers *et al.*, 1983; Amthor, 1989).

Third, there is the problem of the exact N source of the plant. If N is taken up in the form of NH_4^+ exclusively, the plant does not need to provide energy for the reduction step from NO_3^- to NH_4^+. If N is taken up as NO_3^- and reduced in roots or stems, these costs must be included. When a plant uses NO_3^- and reduces this in the leaves, matters are more complicated. Part of the reduction, from nitrite to ammonium, takes place in the chloroplasts, and the reducing power necessary for the reduction could be drawn directly from the NADPH supply by the light reaction (Layzell, 1990). In this way, the steps of converting CO_2 to glucose and respiring it thereafter are avoided. Consequently, these costs do not need to be included in the carbon budget. Most of the studies on construction cost consider NO_3^- as the nitrogen source and assume that nitrate reduction comes at full cost. This will therefore yield maximum values for the construction costs. For a plant with a leaf protein concentration of 230 mg g^{-1}, the difference in construction costs, with NO_3^- or NH_4^+ as nitrogen source, would be 0.20 g glucose g^{-1}. This could represent about 13% of the construction cost. For stems and roots, which have much lower protein concentrations (Fig. 1A), this value will be less. An estimate of the form of N taken up by the plant could be obtained by assessing the difference between cations and anions, relative to the amount of organic N in the total plant (Troelstra, 1983).

Given the assumptions stated above, one may question how well these construction costs are estimated. Not many independent checks have been made. Poorter (1994) determined photosynthesis, respiration, biomass allocation, and chemical composition of a range of herbaceous species, differing in potential RGR by 300%. He calculated the construction costs per species from the chemical composition. With a slight alteration of Eq. (1) he could estimate RGR from these parameters as follows:

$$RGR = \frac{PS_A \times SLA \times LWR - LR'_w \times LWR - SR'_w \times SWR - RR'_w \times RWR}{CC \times \dfrac{6}{180}}, \quad (2)$$

where LR'_w and SR'_w are the maintenance respiration rate for leaves and stem, respectively, RR'_w the respiration rate of the roots related to maintenance and uptake of nutrients, and $6/180$ is the multiplication factor to convert whole plant construction costs (CC) from grams of glucose to moles of C. Maintenance respiration is defined as the respiration an organ needs to supply the energy for keeping the organ functioning. The RGR values for each species calculated in this way were lower than those determined experimentally. Apparently, the costs in terms of CO_2 losses were overestimated to some extent. Given the uncertainties around each of the determinations, this could just be a matter of chance. However, if the assumption that the plants reduced nitrate at the cost of glucose breakdown was lifted, estimated and determined values agreed quite well. This in itself would lend support to the idea that nitrite reduction takes place largely in the chloroplasts, and could consume reducing power which would otherwise not be used for C fixation (Layzell, 1990).

Verification of the construction cost values could proceed at a lower integration level as well. The amount of carbon required for the carbon skeletons of the plant material can be determined precisely. Given a C concentration in plant material of 400 mg g^{-1}, the amount of glucose required to provide the C skeletons for 1 gram of biomass is 1000 mg. The other part of the construction cost is the glucose, which will be respired to provide for the NADPH and ATP required for biosynthesis. This is the so-called growth coefficient, which we express here as the amount of glucose respired to synthesize 1 gram of biomass. It is this part that needs independent confirmation. Plant respiration can be measured relatively easily, but it will include respiration involved in both growth and maintenance. One of the most commonly used methods to separate these two is the regression method developed by Thornley (1970). In this method, total respiration per unit organ weight and time (R) is written:

$$R = m + g \times \mathrm{RGR}, \tag{3}$$

where m and g are the maintenance and growth respiration, respectively, and RGR the specific growth rate of the organ or plant. How well do these independent estimates of g match the amount of glucose that is calculated to form the part of the construction cost which is to be respired? We compiled both g and construction cost values of the literature. We assumed herbaceous and woody plants to have a C concentration of 400 and 450 mg g^{-1}, respectively. By deducting the glucose necessary for C skeletons (1.0 and 1.125 g glucose g^{-1}, respectively, for the two groups of species) from each of the total estimated construction cost values, a second estimate of g was obtained. Somewhat as a surprise, these two independent estimates showed almost exactly the same range and distribution (Fig. 5). This gives some credibility to the approaches used, although we consider the proof still not good enough. Estimates where construction costs and g are determined

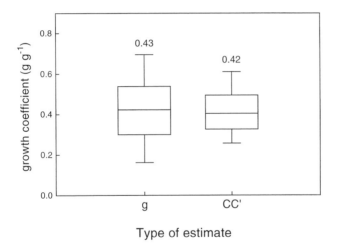

Figure 5 Characterization of the distribution of the growth coefficient g, the respiration required for the construction of 1 gram of biomass, as derived from two independent estimates. In the first approach, g is calculated directly from respiration measurements, using the regression method to separate maintenance and growth respiration. In the second approach (CC'), g is calculated as the difference between total construction costs and the glucose necessary to provide the C skeletons. In this calculation it was assumed that the organs of herbaceous and woody species have a C concentration of 400 and 450 mg g^{-1}, respectively. For comparative purposes, g values are converted from millimoles CO_2 produced per gram of material to gram glucose respired per gram of material formed. Values of g were restricted to vegetative organs and are from Hughes (1973), Merino *et al.* (1982), the values listed in Table 5.1 from Amthor (1989), Baker *et al.* (1992), Wullschleger and Norby (1992), Wullschleger *et al.* (1992), Lehto and Grace (1994), Bunce (1995), Shinano *et al.* (1995), and Wullschleger *et al.* (1995), as well as obtained from D. Garcia (personal communication). Data for construction costs are from the compilation listed in the legend of Fig. 6. Values above the box plots are averages for the category of interest.

independently are scarce. Merino *et al.* (1984) calculated the theoretical *g*, based on chemical composition, for three chaparral species. These values were close (within 7%) to the *g* estimates obtained by the regression method for the same leaves (Merino *et al.*, 1982). More or less similar conclusions could be derived from data by Marcelis and Baan Hofman-Eijer (1995) and Shinano *et al.* (1995).

C. Differences between Organs

Griffin (1994) reviewed the literature on construction costs obtained by calorimetric estimates. He concluded that there is substantial divergence in construction costs between species. However, Poorter (1994), reviewing literature data on construction costs estimates from all available methods, was unable to find systematic differences between functional groups of species and thus concluded the opposite. Since that time, the amount of data has doubled. In this section, we update the review of Poorter (1994). A compilation of these data is given in Fig. 6, with new references listed in the legend.

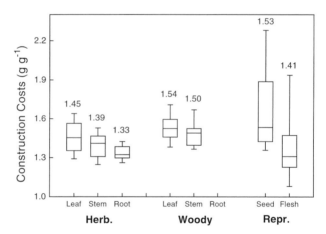

Figure 6 Characterization of the distribution of the construction costs (grams glucose required per gram of biomass formed), by box plots. Data are from the compilation by Poorter (1994) and are supplemented by those of Amthor *et al.* (1994), Sobrado (1994), Walton and Fowke (1995), 76 species from Villar (1992), 27 species (Poorter *et al.*, 1997), 10 species (H. Poorter and J. R. Evans, unpublished), and 60 boreal species (V. I. Pyankov and H. Poorter, unpublished). Furthermore, construction cost values were calculated from Waterman *et al.* (1980), Jordano (1995), and Shinano *et al.* (1995). The total number of observations are as follows: herbs, leaf 157, stem 32, root 32; woody species, leaf 203, stem 26; reproductive organs, seed 47, fruit flesh 294. Values above the box plots are averages for the category of interest.

Most of the research up to now has focused on the construction costs of leaves only. Based on all available data from herbaceous and woody species, we conclude that construction of 1 gram of an average leaf requires 1.50 g glucose (see Fig. 6). There are some data available where stems and roots have been analyzed as well. Generally, stems have somewhat lower construction costs (1.45 g glucose g^{-1}) than leaves. Roots have hardly been investigated, but estimates cluster around 1.33. This could be explained by the difference in chemical composition, as shown in Fig. 1 and Table I. Stems and roots do have lower levels of the expensive compounds lipids and protein, but higher levels of lignin. Concentrations of cheap compounds show differences as well, with organic acids being lower but minerals being higher. Clearly, not just one compound determines the difference. Rather, it is the balance between the various constituents which causes the organs to differ in construction costs. Not much data are present to make inferences on systematic variation in construction costs between organs of woody species. Notwithstanding the high lignin concentrations (Fig. 1A), stems of woody species do not show particularly high construction costs (Fig. 6). This is due to the accompanying large accumulation of TSC.

On average, seeds have similar costs as leaves, although the variation in values is much larger. Especially in seeds with both a high lipid and protein content, construction costs can be over 2 g glucose g^{-1}. Values of the fruit flesh are somewhat lower than those of seeds, although again variation is much larger than in vegetative organs (see Fig. 1).

D. Interspecific Variation

Interspecific variation can be considerable, if reports on individual species are compared. In some cases, values for leaves as low as 1.1 or as high as 2.0 g glucose g^{-1} have been reported. This led Griffin (1994) to the conclusion that there was substantial divergence in construction costs between species. Unfortunately, these reports generally comprise single determinations, so there is no clue as to how consistent these differences are. Furthermore, as a proximate chemical analysis is lacking, there is no clue to the underlying reasons for the low or high values. When construction costs are compared between functional groups of species, differences are small. However, contrary to the previous compilation of Poorter (1994), we detected some systematic interspecific variation. On average, leaves of woody species have 6% higher construction costs than those of herbaceous species (1.54 versus 1.45 g glucose g^{-1}; $P < 0.001$). However, it should be remembered that in this data set all kinds of values are compiled. To what extent are these observations backed up by larger scale comparisons? When we analyze all data on deciduous and evergreen woody species, in reports where both life forms were investigated simultaneously, deciduous species

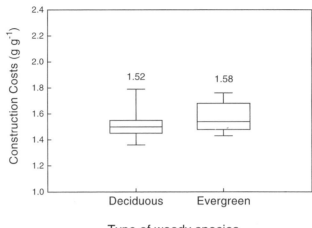

Type of woody species

Figure 7   Characterization of the distribution of construction costs (grams glucose required per gram of biomass formed), by box plots, for leaves of deciduous ($n = 36$) and evergreen ($n = 35$) woody species. Data are from Merino *et al.* (1984), Merino (1987), Chapin (1989), Villar (1992), and Sobrado (1994). Values above the box plots are averages for the category of interest.

show 4% lower leaf construction costs than evergreens ($P < 0.05$; Fig. 7). In a survey of woody species growing across Spain, Merino (1987) found that gymnosperms had higher leaf construction costs than deciduous angiosperms (11% difference) and that evergreen species had higher leaf construction costs than deciduous plants (7% difference). Chapin (1989) did not find systematic differences between arctic plant species of different life forms. In all of these cases variation in environmental parameters may have affected the results. However, a similar difference (11%, $P < 0.01$) between six herbaceous and four woody species was found when plants were grown in growth rooms (H. Poorter and J. R. Evans, unpublished). Therefore, we conclude that there is at least some indication of a small inherent difference in construction costs between leaves of herbaceous and (evergreen) woody species.

Within the group of herbaceous angiosperms, Poorter and Bergkotte (1992) did not find a systematic relationship between leaf and/or plant construction costs and the potential growth rate of 24 species. No other group differences have been explored so far.

The reason for construction costs of vegetative organs being relatively constant is the pattern of covariation between the various classes of constituents (Section IV). In arctic plants, Chapin (1989) observed negative correlations between expensive compounds like protein and lignin, or lignin and tannin. Poorter (1994) concluded that such negative relationships existed in fast- and slow-growing herbaceous species as well, but that quantitatively they could not explain the absence of differences in construction costs. Rather, it was the positive correlation between expensive proteins and cheap minerals that was the main reason for the lack of systematic variation. As noticed before, interspecific variation in construction costs of reproductive organs is much larger (Fig. 6). This is due to the fact that seeds and fruit flesh may contain large concentrations (up to 800 mg g^{-1}) of storage compounds (Fig. 1), which in some cases are costly (lipids, protein) and in other cases are cheap (TNC).

E. Effect of the Environment

Environmental impact on the construction costs is scarcely investigated, with most of the work again restricted to leaves. As a rule, effects are small (Fig. 8). On average, there was a 3% decrease in construction costs with increases in the ambient CO_2 concentration. The decrease is explained by the accumulation of TNC and, independent of that, a decrease in the concentration of protein (Griffin *et al.*, 1993; Poorter *et al.*, 1997). The effect of light is variable. Williams *et al.* (1989) observed both higher and

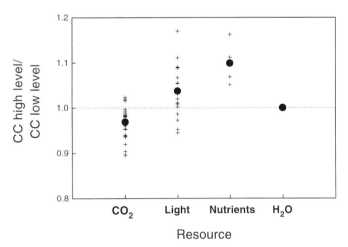

Resource

Figure 8 Differences in leaf construction costs (grams glucose required per gram of biomass formed) due to growth of plants under limiting conditions. Data are from Merino (1987), Lafitte and Loomis (1988), Williams *et al.* (1989), Griffin *et al.* (1993), Amthor *et al.* (1994), Griffin (1994), Sims and Pearcy (1994), Shinano *et al.* (1995), Poorter *et al.* (1997); 27 species grown at 350 and 700 μl liter^{-1} CO_2), and H. Poorter and J. R. Evans (unpublished, 10 species at light intensities of 200 and 1000 μmol quanta m^{-2} s^{-1}). Small pluses indicate ratios observed for leaves of individual species; closed circles indicate the back-transformed average of the ln-transformed individual data. Values above the box plots are averages for the category of interest.

lower construction costs for *Piper* species growing in the field in large versus small gaps. In the controlled experiment outlined above, H. Poorter and J. R. Evans found leaf construction costs to be 4% higher across species for plants grown at a light intensity of 1000 μmol quanta m^{-2}s^{-1} as compared to plants grown at 200 μmol m^{-2}s^{-1}. This scales with the lower concentration of minerals in the leaves of high-light-grown plants.

High levels of N increased construction costs by $\sim$10%. Again, changes in the concentrations of TNC, protein, and minerals will be the main factors, with decreases in the first parameter and increases in the other two. An interesting experiment has been carried out by Peng *et al.* (1993), who grew *Citrus* plants with and without mycorrhiza, at both low and high P. Mycorrhizal roots of high-P plants had lower construction costs than those of high-P plants, which could be ascribed to a lower degree of infection with lipid-rich fungal hyphae. However, it should be noted that also nonmycorrhizal roots at high P had lower construction costs than low-P roots. This is at variance with the increases in leaf costs observed for plants grown at high N availability. Without knowledge of the chemical composition of the *Citrus* roots, it is impossible to assess the reason for this difference in behavior. Controlled experiments with water availability have not been carried out. Merino (1987) reports no differences between plants growing in xeric and in mesic habitats.

VII. Ecological Consequences

Based on the literature available at that time, Poorter (1994) concluded that there was hardly any indication for differences between groups of species. We find an indication that leaves of evergreen shrubs and trees have somewhat higher construction costs than leaves of deciduous species. Moreover, we find higher leaf construction costs for high-light plants, and lower costs for high-CO_2 plants. However, differences are small (well below 10%). Does that mean they are insignificant in relation to growth? Assuming all other parameters are constant, we can assess the effect of a given difference in construction costs on growth with help of Eq. (2). A 10% increase in construction costs translates into a 10% decrease in RGR. In the comparison of fast- and slow-growing species, which may differ in RGR by more than 300% (Grime and Hunt, 1975; Poorter and Remkes, 1990; Garnier, 1992), such a difference is small. However, we cannot exclude the possibility that, in some habitats, such a difference might make for a decisive impact on competition. A similar conclusion may be drawn from the range in construction costs observed for individual species. Given the 27% difference between the 10th and the 90th percentile of the distribution of herbaceous leaf construction costs (Fig. 6), we conclude that in some specific cases the difference in construction costs could lead to a moderate difference in growth.

The costs assessed in this way are the direct costs a plant has to make to construct an organ with a specific chemical composition. However, there are indirect costs and consequences of that as well. If a plant would invest more in protein, it may gear up the photosynthetic and biosynthetic machinery, and in this way increase its growth rate. Alternatively, it may invest more in expensive compounds like lignin and phenolics, which may decrease the risk of herbivory (Bryant *et al.*, 1983; Bazzaz *et al.*, 1987) and increase resistance against decomposition (Enríquez *et al.*, 1993). The relation between chemical composition and these ecologically very important parameters warrants a separate review. We will refrain from that here and only point out that, as a consequence of a high investment in cell-wall compounds and "defense," the proportion of cytoplasmic compounds in the plant decrease, and consequently its rate of C fixation and nutrient acquisition. It will depend on a plant's environment whether such a decreased C acquisition really will confer a cost.

Apart from resistance against herbivory and decomposition, there is one other species attribute that is strongly related to the observed differences in chemical composition: leaf life span. Leaf life span is an important ecological factor in itself, which may affect the outcome of competition (Berendse and Elberse, 1989). Leaf longevity is high for species with a suite of traits characteristic of species from nutrient-poor environments: a low specific leaf area, a low water content per unit dry weight, and a high investment in cell-wall compounds (Lambers and Poorter, 1992; Reich *et al.*, 1992). Contrary to the mechanistic insights into herbivory and decomposition, we have no clue to the mechanisms that determine leaf (and fine root) longevity. Establishing the relative role of morphology, chemical composition, and the genetic program would be one of the areas where ecophysiology could contribute to ecological insights.

VIII. Summary

In this chapter we analyzed the chemical composition of plants by categorizing compounds into eight distinct classes: lipids, soluble phenolics, protein, lignin, total structural carbohydrates (TSC), total nonstructural carbohydrates (TNC), organic acids, and minerals. First, we assessed the concentrations of these compounds in leaves, stems, and roots of herbaceous and woody species, and seeds and fruit flesh. Concentrations of lipids, organic acids, soluble phenolics, and protein are higher in leaves, whereas TSC, lignin, and TNC are generally higher in stems and roots. Woody species have lower concentrations of protein, minerals, and organic acids, and higher levels of soluble phenolics.

Concentrations of the different compounds do not vary independently of one another. We investigated the patterns that emerge from two larger scale comparisons across species. Some species have high concentrations of protein, minerals, as well as organic acids (cytoplasmic plus vacuolar compounds), whereas others have relatively high concentrations of lignin and TSC (cell-wall compounds). These patterns coincide with the water content of the plant material, the leaf area : leaf weight ratio (specific leaf area), as well as the potential relative growth rate of these species. We discussed a number of mechanistic interpretations for these results but concluded that there is insufficient insight into the regulation of the chemical composition.

Depending on their biosynthetic pathway, the various classes of compounds require different amounts of glucose to provide for C skeletons, reducing power, and energy. The total amount of glucose required to construct 1 gram of a compound yields the construction costs of that compound. These construction costs vary 3-fold for organic compounds, and are basically nil for the minerals. Consequently, construction costs for 1 gram of biomass will depend on its chemical composition. Although large differences in chemical composition are found among species or between plants grown in different environments, differences in construction costs of biomass are generally small (up to 10%) or nonexistent. The positive correlation between expensive and cheap compounds (proteins and minerals/organic acids) or the negative correlation between various expensive ones (protein and lignin) explains the relatively stable construction costs when groups of functional types of species are compared. However, there are some reports on relatively high or low construction costs for leaves of specific species, which warrant more attention. The same holds in general for stems and roots, which are hardly investigated.

IX. Appendix 1

Short Overview of the Chemical Determinations

This is a short overview of the methods currently used in the Utrecht University laboratory to determine the different fractions of plant compounds (see also Fig. 9). Especially with respect to sugars and phenolic compounds, it is advisable to use freeze-dried material. To avoid too many determinations and to ensure sufficient biomass, we pool (for each organ separately) dried biomass from different individuals, to arrive at two independent samples per species, treatment and harvest. Each of these samples is ground to pass through a 0.08-mm sieve. Total dry weight (DW) of each

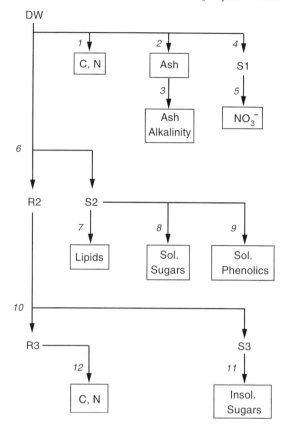

Figure 9 Flowchart of a convenient procedure for determination of the proximate chemical composition. A description is given in Appendix 1.

of the subsamples after redrying should be ~750 mg. This will allow each of the subsequent determinations to be carried out in duplicate. Calculations are discussed in Appendix 2.

Total C and N are determined with an elemental analyzer, which requires 1 mg of material per sample (*1*, numbers refer to Fig. 9). A second fraction, at least 100 mg, is weighed, ashed at 550°C in a muffle furnace, and weighed again (*2*). The ash consists partly of minerals, partly of oxides that are derived from organic acids, and nitrate. These compounds disappear during ashing except for O^{2-} (Dijkshoorn *et al.*, 1968). On cooling the oxides react with CO_2 to form CO_3^{2-}. The total amount of carbonates is determined by quantitatively transferring the ash in an Erlenmeyer flask and determining ash alkalinity by first adding 0.05 N HCl and thereafter titrating back with 0.05 N NaOH (*3*). Before being able to calculate total mineral concentration and organic acid concentrations, NO_3^- has to be determined as well. Approximately 25 mg per sample is weighed, and the nitrate is extracted with water at 80°C (*4*). Subsequently, concentrations are determined in the supernatant S1 following Cataldo *et al.* (1975; *5*).

The above determinations are enough to arrive at an estimate of the construction costs. If the aim is to determine the proximate composition of the plant material, then another sample of the ground and redried biomass (±250 mg) is extracted with a mixture of chloroform/methanol according to Bligh and Dyer (1959; *6*). Addition of water produces a chloroform phase and a methanol/water phase. The chloroform is evaporated with N_2 and the residue weighed. This residue largely contains phospholipids and galactolipids, as well as some sterols, and is termed lipids in this chapter (*7*). The water/methanol phase contains the soluble sugars (glucose, sucrose, soluble fructan, etc.), which can be determined with the anthrone method (*8*; Fales, 1951). In the same phase are the soluble phenolics, which can be measured with the Folin–Ciocalteu reagent (*9*).

The residue R2, mainly cell debris left over after extraction with chloroform/methanol, is boiled for 3 hours at 100°C with 3% HCl. This will break down starch as well as the remainder of the fructans, pectins, and some part of the hemicellulose (*10*). The sugars released on acid hydrolysis are determined with the anthrone method (*11*). The residue that is left over after this extraction, R3, consists of cellulose, hemicellulose, cell wall protein, protein that had precipitated during the first extraction, and lignin. After drying and weighing this residue is pulverized, and a sample is taken for a C and N analysis in the elemental analyzer (*12*).

X. Appendix 2

Calculations

The concentrations of lipid, soluble phenolics, and soluble and insoluble sugars are determined directly either by weighing or with calibration curves. Protein is calculated by subtracting NO_3^- N from total N and multiplying this difference by 6.25. Organic acid concentration (OA) is derived from ash alkalinity (AA), expressed in mEq g^{-1}, and the NO_3^- concentration, also expressed in mEq g^{-1}:

$$OA = \frac{AA - Nit}{62}, \tag{A1}$$

where 62 is the weight of 1 Eq of organic acids. This number 62 is the average value determined by gas chromatography for some grasses grown hydroponically and may be different in cases where malate is not the main organic acid.

The mineral concentration (Min) is calculated as

$$\text{Min} = \text{Ash} - \text{AA} \times 30 + \text{Nit}, \tag{A2}$$

where Ash and Nit are expressed as mg g^{-1} dry weight, and ash alkalinity (AA) expressed in mEq g^{-1} is multiplied by the weight of carbonate per equivalent of charge.

When the concentrations of C, organic N, and minerals (Min; all in mg g^{-1}) are known, the construction costs (in g glucose g^{-1}) can be calculated (Vertregt and Penning de Vries, 1987; modified by Poorter, 1994) as follows:

$$CC = \left(-1.041 + 5.077 \times \frac{C}{1000 - \text{Min}} \right) \times \frac{1000 - \text{Min}}{1000} + \left(5.325 \times \frac{N_{\text{org}}}{1000} \right). \tag{A3}$$

The available methods to determine lignin are generally rough, use very aggressive chemicals, show interference with other plant compounds, and have to be calibrated with standards (e.g., Morrison, 1972; Morrison et al., 1995). As no method currently exists to obtain pure lignin, these standards are not precise either and generally consist of a gravimetrically determined residue of plant material treated with a series of increasingly aggressive chemicals. Another drawback of this method is that calibrations actually have to be carried out for each species separately, which makes comparative analyses extremely tedious. Therefore, we chose to calculate lignin concentrations in a quicker and easier way, which, though not fully precise either, avoids a number of the above problems. The essence of this calculation is that lignin has an almost 50% higher C concentration than the TSC complex. Starting from the C concentration of R3 (C_{R3}; see Fig. 9), we can arrive at an estimate of the lignin concentration. Given the total weight of the protein fraction, and the concentration of C in plant protein in general (C_{prot}; 530 mg g^{-1}) we can calculate the C concentration of the nonprotein fraction of the residue, C':

$$C' = \frac{C_{R3} \times W_{R3} - C_{\text{prot}} \times N_{R3} \times 6.25}{W'_{R3}} \tag{A4}$$

where W_{R3} is the weight of residue R3. The weight of the nonprotein fraction, W'_{R3}, is given by

$$W'_{R3} = W_{R3} - N_{R3} \times 6.25. \tag{A5}$$

We assume that this fraction consists of two compounds only: TSC (W_{TSC}) and lignin (W_{Lig}):

$$W'_{R3} = W_{\text{TSC}} + W_{\text{Lig}}. \tag{A6}$$

The C concentration of this fraction will depend on the relative contribution of TSC (with a C concentration C_{TSC}) and the lignin fraction (with a C concentration C_{Lig}):

$$C' = \frac{W_{TSC}}{W'_{R3}} \times C_{TSC} + \frac{W_{Lig}}{W'_{R3}} \times C_{Lig} \tag{A7}$$

The TSC complex and lignin both have variable C contents, depending on the relative contribution of their building blocks. However, values of 440 and 640 mg g^{-1} (Fengel and Wegener, 1989) are reasonable approximations to work with. Then, there are two equations [Eqs. (A6) and (A7)] with two unknowns. The lignin concentration in the pellet can be calculated as

$$\frac{W_{Lig}}{W'_{R3}} = \frac{C' - C_{TSC}}{C_{Lig} - C_{TSC}}. \tag{A8}$$

TSC forms the remainder. Given the concentrations in the pellet, concentrations of lignin and TSC in the total dry weight sample can be readily calculated.

Acknowledgments

We thank Yvonne van Berkel and Robin Liebrechts for help with the chemical determinations. Pedro Jordano, Vladimir Pyankov, and Graeme Batten kindly made part of their data available. Marion Cambridge, Eric Garnier, Henri Groeneveld, Hans Lambers, Esther Perez-Corona, Lourens Poorter, Robbert Scheffer, and Adrie van der Werf made valuable comments on some of the previous versions of this paper. RV acknowledges financial assistance by the Spanish DGICYT Project PB92-0813.

References

Amthor, J. S. (1989). "Respiration and Crop Productivity." Springer-Verlag, Berlin.
Amthor, J. S., Mitchell, R. J., Runion, G. B., Rogers, H. H., Prior, S. A., and Wood, C. W. (1994). Energy content, construction cost and phytomass accumulation of *Glycine max* (L.) Merr. and *Sorghum bicolor* (L.) Moench grown in elevated CO_2 in the field. *New Phytol.* **128**, 443–450.
Baker, J. T., Laugel, F., Boote, K. J., and Allen, H. (1992). Effects of daytime carbon dioxide concentration on dark respiration in rice. *Plant Cell Environ.* **15**, 231–239.
Batten, G. D., Blakeney, A. B., McGrath, V. B., and Ciavarella, S. (1993). Non-structural carbohydrate: Analysis by near infrared reflectance spectroscopy and its importance as an indicator of plant growth. *Plant Soil* **156**, 243–246.
Bazzaz, F. A., Chiariello, N. R., Coley, P. D., and Pitelka, L. F. (1987). Allocating resources to reproduction and defense. *BioScience* **37**, 58–67.
Berendse, F., and Elberse, W.T. (1989). Competition and nutrient losses from the plant. *In* "Causes and Consequences of Variation in Growth Rate and Productivity of Higher Plants" (H. Lambers, M. L. Cambridge, H. Konings, and T. L. Pons, eds.), pp. 269–284. SPB Academic Publ., The Hague.

Berg, B., and Staaf, H. (1980). Decomposition rate and chemical changes of Scots pine litter. II. Influence of chemical composition. *Ecol. Bull.* **32**, 373–390.

Bligh, E. G., and Dyer, W. J. (1959). A rapid method of total lipid extraction and purification. *Can. J. Biochem. Physiol.* **37**, 911–917.

Bliss, L. C. (1962). Caloric and lipid content in alpine tundra plants. *Ecology* **43**, 753–755.

Blom-Zandstra, M., Lampe, J. E. M., and Ammerlaan, H. M. (1988). C and N utilization of two lettuce genotypes during growth under non-varying light conditions and after changing the light intensity. *Physiol. Plant.* **74**, 147–153.

Boon, J. J. (1989). An introduction to pyrolysis mass spectrometry of lignocellulosic material: Case studies on barley straw, corn stem and *Agropyron*. *In* "Physicochemical Characterization of Plant Residues for Industrial and Feed Use" (A. Chesson and E. R. Ørskov, eds.), pp. 25–45. Elsevier, Amsterdam.

Bryant, J. P., Chapin, F. S., and Klein, D. R. (1983). Carbon/nutrient balance of boreal plants in relation to vertebrate herbivory. *Oikos* **40**, 357–368.

Buckingham, J. (1993). "Dictionary of Natural Products." Chapman & Hall, London.

Bunce, J. A. (1995). The effect of carbon dioxide concentration on respiration of growing and mature soybean leaves. *Plant Cell Environ.* **18**, 575–581.

Caspers, N. (1977). Seasonal variation of caloric values in herbaceous plants. *Oecologia* **26**, 379–383.

Cataldo, D. A., Haroon, M., Schrader, L. E., and Youngs, V. (1975). Rapid colorimetric determination of nitrate in plant tissue by nitration of salicylic acid. *Commun. Soil Sci. Plant Anal.* **6**, 71–80.

Challa, H. (1976). "An Analysis of the Diurnal Course of Growth, Carbon Dioxide Exchange and Carbohydrate Reserve Content of Cucumber." Agricultural Research Report 861. Pudoc, Wageningen.

Chapin, F. S. (1989). The costs of tundra plant structures: Evaluation of concepts and currencies. *Am. Nat.* **133**, 1–19.

Chiariello, N. R., Mooney, H. A., and Williams, K. (1989). Growth, carbon allocation and cost of plant tissue. *In* "Plant Physiological Ecology: Field Methods and Instrumentation" (R. W. Pearcy, J. R. Ehleringer, H. A. Mooney, and P. W. Rundel, eds.), pp. 327–365. Chapman & Hall, London.

Dijkshoorn, W., Lathwell, D. J., and De Wit, C. T. (1968). Temporal changes in carboxylate content of ryegrass with stepwise change in nutrition. *Plant Soil* **29**, 369–390.

Dijkstra, P. (1989). Cause and effect of differences in specific leaf area. *In* "Causes and Consequences of Variation in Growth Rate and Productivity of Higher Plants" (H. Lambers, M. L. Cambridge, H. Konings, and T. L. Pons, eds.), pp. 125–140. SPB Academic Publ., The Hague.

Duke, S. A., and Atchley, A. A. (1986). "CRC Handbook of Proximate Analysis Tables of Higher Plants." CRC Press, Boca Raton, Florida.

Enríquez, S., Duarte, C.M., and Sand-Jensen, K. (1993). Patterns in decomposition rates among photosynthetic organisms: The importance of detritus C:N:P content. *Oecologia* **94**, 457–471.

Fales, F.W. (1951). The assimilation and degradation of carbohydrates by yeast cells. *J. Biol. Chem.* **193**, 113–124.

Fengel, D., and Wegener, G. (1989). "Wood. Chemistry, Ultrastructure, Reactions." de Gruyter, Berlin.

Garnier, E. (1992). Growth analysis of congeneric annual and perennial grass species. *J. Ecol.* **80**, 665–675.

Garnier, E., and Laurent, G. (1994). Leaf anatomy, specific mass and water content in congeneric annual and perennial grass species. *New Phytol.* **128**, 725–736.

Gojon, A., Plassard, C., and Bussi, C. (1994). Root/shoot distribution of NO_3^- assimilation in herbaceous and woody species. *In* "A Whole Plant Perspective on Carbon–Nitrogen Interactions" (J. Roy and E. Garnier, eds.), pp. 131–147. SPB Academic Publ., The Hague.

Griffin, K. L. (1994). Calorimetric estimates of construction costs and their use in ecological studies. *Funct. Ecol.* **8**, 551–562.

Griffin, K. L., Thomas, R. B., and Strain, B. R. (1993). Effects of nitrogen supply and elevated carbon dioxide on construction costs in leaves of *Pinus taeda* (L.) seedlings. *Oecologia* **95**, 575–580.

Grime, J. P., and Hunt, R. (1975). Relative growth rate: Its range and significance in a local flora. *J. Ecol.* **63**, 393–422.

Hehl, G., and Mengel, K. (1972). Der Einfluss einer variierten Kalium- und Stickstoffdüngung auf den Kohlenhydratgehalt verschiedener Futterpflanzen. *Landwirtsch. Forsch. Sonderh.* **27**, 117–129.

Hughes, A. P. (1973). A comparison of the effects of light intensity and duration on *Chrysanthemum morifolium* cv. Bright Golden Anne in controlled environments. *Ann. Bot.* **37**, 275–280.

Jordano, P. (1995). Angiosperms fleshy fruits and seed dispersers: A comparative analysis of adaptation and constraints in plant–animal interactions. *Am. Nat.* **145**, 163–191.

Kedrowski, R. A. (1983). Extraction and analysis of nitrogen, phosphorus and carbon fractions in plant material. *J. Plant Nutr.* **6**, 989–1011.

Körner, C., Pelaez-Riedl, S., and Van Bel, A.J.E. (1995). CO_2 responsiveness of plants: A possible link to phloem loading. *Plant Cell Environ.* **18**, 595–600.

Lafitte, H.R., and Loomis, R.S. (1988). Calculation of growth yield, growth respiration and heat content of grain sorghum from elemental and proximal analysis. *Ann. Bot.* **62**, 353–361.

Lambers, H. (1985). Respiration in intact plants and tissues: Its regulation and dependance on environmental factors, metabolism and invaded organisms. *In* "Encyclopaedia of Plant Physiology, New Series, Volume 18" (R. Douce and D. Day, eds.), pp. 418–473. Springer-Verlag, Berlin.

Lambers, H. (1993). Rising CO_2, secondary plant metabolism, plant herbivore interactions and litter composition. *Vegetatio* **104/105**, 263–271.

Lambers, H., and Poorter, H. (1992). Inherent variation in growth rate between higher plants: A search for physiological causes and ecological consequences. *Adv. Ecol. Res.* **23**, 188–261.

Lambers, H., and Rychter, A. (1989). The biochemical background of variation in respiration rate: Respiratory pathways and chemical composition. *In* "Causes and Consequences of Variation in Growth Rate and Productivity of Higher Plants" (H. Lambers, M. L. Cambridge, H. Konings, and T. L. Pons, eds.), pp. 199–225. SPB Academic Publ., The Hague.

Lambers, H., Szaniawski, R. K., and De Visser, R. (1983). Respiration for growth, maintenance and ion uptake. An evaluation of concepts, methods, values and their significance. *Physiol. Plant.* **58**, 556–563.

Landsberg, J. (1987). Feeding preferences of common Brushtails Possums, *Trichosurus vulpecula*, on seedlings of a woodland eucalypt. *Aust. Wildl. Res.* **14**, 361–369.

Layzell, D.B. (1990). N_2 fixation, NO_3 reduction and NH_4 assimilation. *In* "Plant Physiology, Biochemistry and Molecular Biology" (D. T. Dennis and D. H. Turpin, eds.), pp. 389–406. Longman Group U.K., Harlow.

Lehto, T., and Grace, J. (1994). Carbon balance of tropical tree seedlings: A comparison of two species. *New Phytol.* **127**, 455–463.

Loveless, A. R. (1962). Further evidence to support a nutritional interpretation of sclerophylly. *Ann. Bot.* **26**, 551–561.

Marcelis, L. F. M., and Baan Hofman-Eijer, L. R. (1995). Growth and maintenance respiratory costs of cucumber fruits as affected by temperature, and ontogeny and size of fruits. *Physiol. Plant.* **93**, 484–492.

Margna, U. (1977). Control at the level of substrate supply—An alternative in the regulation of phenolpropanoid accumulation in plant cells. *Phytochemistry* **16**, 419–426.

McDonald, A. J. S., Ericsson, A., and Lohammer, T. (1986). Dependence of starch storage on nutrient availability and photon flux density in small birch (*Betula pendula* Roth). *Plant Cell Environ.* **9**, 433–438.

Melillo, J. M., Aber, J. D., and Muratore, J. E. (1982). Nitrogen and lignin control of hardwood leaf litter decomposition dynamics. *Ecology* **63**, 621–626.

Merino, J. (1987). The costs of growing and maintaining leaves of mediterranean plants. *In* "Plant Response to Stress" (J. D. Tenhunen, F. M. Catarino, O. L. Lange, and W. C. Oechel, eds.), pp. 553–564. Springer-Verlag, Berlin.

Merino, J., Field, C., and Mooney, H. A. (1982). Construction and maintenance costs of mediterranean-climate evergreen and deciduous leaves. I. Growth and CO$_2$ exchange analysis. *Oecologia* **53**, 211–229.

Merino, J., Field, C., and Mooney, H.A. (1984). Construction and maintenance costs of mediterranean-climate evergreen and deciduous leaves. *Acta Oecol./Oecol. Plant.* **5**, 211–229.

Mooney, H. A., Fichtner, K., and Schulze, E.D. (1995). Growth, photosynthesis and storage of carbohydrates and nitrogen in *Phaseolus lunatus* in relation to resource availability. *Oecologia* **104**, 17–23.

Morrison, I. M. (1972). A semi-micro method for the determination of lignin and its use in predicting the digestibility of forage crops. *J. Sci. Food Agric.* **23**, 455–463.

Morrison, I. M., Asiedu, E. A., Stuchbury, T., and Powell, A. A. (1995). Determination of lignin and tannin contents of cowpea seed coats. *Ann. Bot.* **76**, 287–290.

Niemann, G. J., Pureveen, J. B. M., Eijkel, G. B., Poorter, H., and Boon, J. J. (1992). Differences in relative growth rate in 11 grasses correlate with differences in chemical composition as determined by pyrolysis mass spectrometry. *Oecologia* **89**, 567–573.

Niemann, G. J., Eijkel, E. B., Konings, H., Pureveen, J. B. M., and Boon, J. J. (1993). Chemical differences between wild-type and gibberellin mutants of tomato determined by pyrolysis-mass spectrometry. *Plant Cell Environ.* **16**, 1059–1069.

Peng, S., Eissenstat, D. M., Graham, J. H., Williams, K., and Hodge, N. C. (1993). Growth depression in mycorrhizal citrus at high phosphorus supply. Analysis of carbon costs. *Plant Physiol.* **101**, 1063–1071.

Penning de Vries, F. W. T., Brunsting, A. H. M., and Van Laar, H. H. (1974). Products, requirements and efficiency of biosynthetic processes: a quantitative approach. *J. Theor. Biol.* **45**, 339–377.

Penning de Vries, F. W. T., Van Laar, H. H., and Chardon, M. C. M. (1983). Bioenergetics of growth of seeds, fruits, and storage organs. *In* "Proceedings of the Symposium on Potential Productivity of Field Crops under Different Environments, 1980," pp. 37–59. International Rice Institute, Manila, Philippines.

Poorter, H. (1989). Interspecific variation in relative growth rate: On ecological causes and physiological consequences. *In* "Causes and Consequences of Variation in Growth Rate and Productivity of Higher Plants" (H. Lambers, M. L. Cambridge, H. Konings, and T. L. Pons, eds.), pp. 45–68. SPB Academic Publ., The Hague.

Poorter, H. (1994). Construction costs and payback time of biomass: A whole plant perspective. *In* "A Whole Plant Perspective on Carbon–Nitrogen Interactions" (J. Roy and E. Garnier, eds.),pp. 111–127. SPB Academic Publ., The Hague.

Poorter, H., and Bergkotte, M. (1992). Chemical composition of 24 wild species differing in relative growth rate. *Plant Cell Environ.* **15**, 221–229.

Poorter, H., and Remkes, C. (1990). Leaf area ratio and net assimilation rate of 24 wild species differing in relative growth rate. *Oecologia* **83**, 553–559.

Poorter, H., Van Berkel, Y., Baxter, B., Bel, M., Den Hertog, J., Dijkstra, P., Gifford, R. M., Griffin, K. L., Roumet, C., and Wong, S. C. (1997). The effect of elevated CO$_2$ on chemical composition and construction costs of leaves of 27 species. *Plant Cell Environ.*, in press.

Raven, J. A. (1985). Regulation of pH and generation of osmolarity in vascular plants: A cost–benefit analysis in relation to efficiency of use of energy, nitrogen and water. *New Phytol.* **101**, 25–77.

Reich, P. B., Walters, M. B., and Ellsworth, D. S. (1992). Leaf life-span in relation to leaf, plant, and stand characteristics among diverse ecosystems. *Ecol. Monogr.* **62**, 365–392.

Shinano, T., Osaki, M., and Tadano, T. (1995). Comparison of growth efficiency between rice and soybean at the vegetative growth stage. *Soil. Sci. Plant Nutr.* **41**, 471–480.

Sims, D., and Pearcy, R. W. (1994). Scaling sun and shade photosynthetic acclimation of *Alocasia macrorrhiza* to whole-plant performance. I. Carbon balance and allocation at different daily photon flux densities. *Plant Cell Environ.* **17**, 881–887.

Sobrado, M. A. (1994). Leaf age effects on photosynthetic rate, transpiration rate and nitrogen content in a tropical dry forest. *Physiol. Plant.* **90**, 210–215.

Thornley, J. H. M. (1970). Respiration, growth and maintenance in plants. *Nature (London)* **227**, 304–305.

Troelstra, S. P. (1983). Growth of *Plantago lanceolata* and *Plantago major* on a NO_3/NH_4 medium and the estimation of the utilization of nitrate and ammonium from ionic-balance aspects. *Plant Soil* **70**, 183–197.

Van Arendonk, J. J. C. M., and Poorter, H. (1994). The chemical composition and anatomical structure of leaves of grass species differing in relative growth rate. *Plant Cell Environ.* **17**, 963–970.

Van Bel, A. J. E. (1994). Strategies of phloem loading. *Annu. Rev. Plant Physiol. Plant Mol. Biol.* **44**, 253–281.

Van der Werf, A., Poorter, H., and Lambers, H. (1994). Respiration as dependent on a species' inherent growth rate and on the nitrogen supply to the plant. *In* "A Whole Plant Perspective on Carbon–Nitrogen Interactions" (J. Roy and E. Garnier, eds.), pp. 91–110. SPB Academic Publ., The Hague.

Vertregt, N., and Penning de Vries, F. W. T. (1987). A rapid method for determining the efficiency of biosynthesis of plant biomass. *J. Theor. Biol.* **128**, 109–119.

Villar, R. (1992). Costos energéticos en hojas de especies leñosas. Ph.D. thesis. Universidad de Sevilla, Spain.

Walton, E. F., and Fowke, P. J. (1995). Estimation of the annual cost of kiwi fruit vine growth and maintenance. *Ann. Bot.* **76**, 617–623.

Waring, R. H., McDonald, A. J. S., Larsson, T., Wiren, A., Arwidsson, E., Ericsson, A., and Lohammer, T. (1985). Differences in chemical composition of plants grown at constant relative growth rates with stable mineral nutrition. *Oecologia* **66**, 157–160.

Waterman, P. G., Mbi, C. N., McKey, D. B., and Gartlan, J. S. (1980). African rainforest vegetation and rumen microbes: Phenolic compounds and nutrients as correlates of digestibility. *Oecologia* **47**, 22–33.

Wedler, A. (1980). Untersuchingen über Nitratgehalte in einigen ausgewählten Gemüsearten. *Landwirtsch. Forsch. Sonderh.* **36**, 128–137.

Williams, K., Percival, F., Merino, J., and Mooney, H. A. (1987). Estimation of tissue construction cost from heat of combustion and organic nitrogen content. *Plant Cell Environ.* **10**, 725–734.

Williams, K., Field, C. B., and Mooney, H.A. (1989). Relationships among leaf construction cost, leaf longevity, and light environment in rain forest plants of the genus *Piper. Am. Nat.* **133**, 198–211.

Wong, S. C., Kriedemann, P. E., and Farquhar, G. D. (1992). $CO_2 \times$ nitrogen interaction on seedling growth of four species of eucalypt. *Aust. J. Bot.* **40**, 457–472.

Wullschleger, S. D., and Norby, R. J. (1992). Respiratory cost of leaf growth and maintenance in white oak saplings exposed to atmospheric CO_2 enrichment. *Can. J. For. Res.* **22**, 1717–1721.

Wullschleger, S. D., Norby, R. J., and Gunderson, C. A. (1992). Growth and maintenance respiration in leaves of *Liriodendron tulipifera* L. exposed to long-term carbon dioxide enrichment in the field. *New Phytol.* **121**, 515–523.

Wullschleger, S. D., Norby, R. J., and Hanson, P. J. (1995). Growth and maintenance respiration in stems of *Quercus alba* after four years of CO_2 enrichment. *Physiol. Plant.* **93**, 47–54.

3

Resource Allocation in Variable Environments: Comparing Insects and Plants

Carol L. Boggs

I. Introduction

Environments are often variable; resources and physical conditions fluctuate across space and time. Environmental variation can affect the life history and foraging behavior of individuals, which in turn may affect population and community dynamics (Fig. 1). We can explore the effects of environmental variation through the use of individual-based models of population dynamics. These models include variation among individuals and age classes in life history parameters, variation that results from a variable environment. This general approach is being used for a few case study organisms, such as *Daphnia* (e.g., Gatto *et al.*, 1989; Koojiman *et al.*, 1989; Gurney *et al.*, 1990; Hallam *et al.*, 1990; McCauley *et al.*, 1990). In the long run, these mechanistic models are powerful tools for understanding population fluctuations in response to natural environmental variation, or to anthropogenic environmental changes ranging from livestock grazing to flood control measures to global climate change.

In order to build such models, however, we need broader generalizations concerning both individual and age class responses to environmental variation. We must understand both the genetics governing the response to environmental variation, and the physiology and ecology of the response. Genotypic constraints limit the phenotypic plasticity exhibited by individuals in response to fluctuations in the environment (e.g., de Jong and van Noordwijk, 1992). These underlying genetic constraints may be subject to

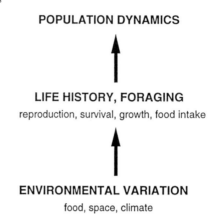

Figure 1   Effects of environmental variation on individuals and populations.

change over evolutionary time. Physiological or ecological aspects of the response to environmental variation are important over shorter time scales, and are reflected in allocation of time and resources in response to variable environments.

Here I focus on general hypotheses concerning how altered resource allocation in response to environmental variation affects an individual's foraging and life history traits. These hypotheses treat changes in allocation among different age classes or developmental stages within an individual, among individuals within a population, and among individuals in different populations within a species. Tests of these allocation hypotheses are drawn from both my work on insects and others' work on plants. Similarities and differences in the biology of these two groups allow a broader test of the general applicability of the ideas. Similarities between insects and plants include the following:

Insects and plants rely on networked gas exchange systems. Insects have a tracheolar system that eliminates any need to carry gases in the hemolymph, while plants have a similar stomatal system.

Both groups primarily thermoregulate behaviorally through orientation to sunlight, although both contain species capable of more active thermogenesis.

Both insects and plants can include resource uptake by the female as part of the reproductive process. Mating in several orders of insects includes donations of nutrients from the male to the female (Thornhill, 1976; Boggs and Gilbert, 1979; Boggs, 1995); seed capsules or flower parts in some plant species have photosynthetic capabilities (e.g., Bazzaz *et al.*, 1979; Galen *et al.*, 1993).

Both groups can manufacture defensive compounds, so that allocation can play an active role in predator escape.

The two groups differ, however, in several ways:

Most insects are heterotrophic, whereas most plants are autotrophic.

Most insects are mobile, whereas many plants can move only slowly in response to environmental conditions by sending out rhizomes in specific directions (Bazzaz, 1991).

Carbohydrate storage in insects takes the form of glycogen, whereas starch is used in plants.

Insects cannot employ indeterminant development because of constraints imposed by the cuticular exoskeleton, but this form of development within limits is common in plants.

II. Background: Resource Allocation within an Individual

A generalized resource flow diagram for an individual is shown in Fig. 2 (see Boggs, 1992). Resources available in the environment are taken into

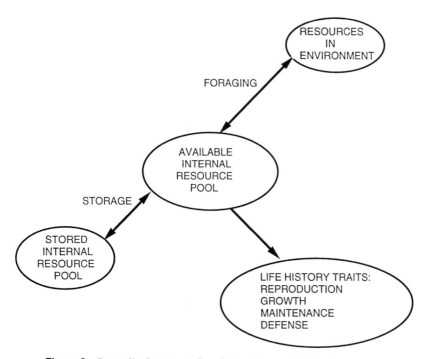

Figure 2 Generalized resource flow diagram for an individual organism.

the organism through foraging; resources absorbed into the body are then allocated to various life history functions, including reproduction, survival/ maintenance, defense, and growth, as well as to storage or support of further foraging for more resources. These life history and foraging traits can be viewed as resource sinks, pulling resources out of the available pool of resources, with each trait's sink strength depending on demand for resources to be used for that particular trait.

Environmental variation can be superimposed on the flow diagram (Fig. 2) in two ways. First, resource availability could vary over either space or time. This may result in either altered resource intake or altered allocation to foraging. If food becomes more dense and easier to find, for example, fewer resources need to be allocated to foraging to maintain a consistent incoming resource flow. Second, expenditure needs may vary over time. For example, reproductive opportunities may not be evenly spread over the life span, metabolic demand may fluctuate with temperature, or needs for defense may vary with size. This will result in temporal variation in the sink strength of each life history trait, with consequent effects on allocation to other traits, to storage, or to foraging. Environmental variation thus may restrict or enhance both inputs and outputs to the resource flow diagram.

In response to environmental variation, organisms may show trade-offs in allocation to life history and foraging traits. Trade-offs between reproduction and survival, or growth and survival, or reproduction and foraging, for example, could occur. These types of trade-offs are widely postulated to result from increasing sink strength (e.g., for reproduction), with the assumption that intake remains constant (e.g., Gadgil and Bossert, 1970; Stearns, 1976). Likewise, these trade-offs are expected if resource intake drops. Such trade-offs that are dependent on the size of the available resource pool are distinct from trade-offs between, for example, reproduction and survival that are due to accumulation of damage to somatic tissue during reproduction, which results in decreased survival (e.g., Partridge and Farquhar, 1981; Partridge, 1987; Tatar *et al.*, 1993).

An alternative response to environmental variation is a change in allocation to foraging, resulting in changes in resource intake (Tuomi *et al.*, 1983; Abrams, 1991; Boggs, 1992). This may eliminate life history and foraging trade-offs. Consider a case where environmental variation results in increasing sink strength for life history traits such as reproduction. An increase in allocation to foraging can lead to an increase in resource intake, resulting in a larger pool of resources that meets the needs for reproduction without decreasing allocation to other traits. This scenario is possible whenever an increase in foraging effort can yield a net increase in resource intake. Alternatively, consider a case where resources in the environment become more costly to obtain. If foraging effort stays constant, then intake is reduced, and total allocation to life history traits will be reduced. However, increasing foraging effort may result in a net increase in resource intake that maintains the available resource pool. If so, allocation to life history traits stays constant, and no trade-offs occur. This scenario again requires that a net increase in resource intake can result from increasing foraging effort.

In all of these cases, organisms must exhibit *allocation plasticity*, or the ability to shift resource flow among different categories of use, altering the rates of allocation to different sinks or traits. Allocation plasticity will depend on (1) the particular types of compounds (e.g., proteins, carbohydrates) needed for each trait, along with the availability and the interconvertibility of those compounds, (2) the metabolic pathways used in allocating resources to each trait, and (3) enzyme variants in those pathways, controlling pathway dynamics. The ability of an organism to exhibit a plastic response will also depend on the size and speed of allocation changes possible, relative to the size and speed of environmental variation (Bazzaz, 1996). That is, if the environment is changing over a time scale of minutes, but the organism can only respond over a time scale of hours, what plastic response occurs can only be an integration over a series of environmental changes.

III. Age-Specific Allocation Responses to Variable Environments

A. Hypothesis

The ability to alter body architecture enhances allocation plasticity, optimizing each age group's allocation patterns.

Body architecture often constrains resource allocation to life history and foraging traits. For example, beak size in Darwin's finches affects the cost of eating (or the possibility of eating) seeds of particular size (Grant *et al.*, 1976). Body cavity size constrains the amount of reproduction in many animals (see Calder, 1984, for a thorough review). "Furriness," or the depth of modified scales on the adult thorax, affects temperature regulation in *Colias* butterflies, with consequent effects on available flight time during which reproduction and foraging occur (Kingsolver, 1983a,b; Jacobs and Watt, 1994).

Changes in body architecture are a form of allocation plasticity. Such changes should result in shifts to new combinations of life history and foraging patterns, as resource allocation changes under the new architectural constraints. Architectural changes are predicted to allow an organism to optimize structure, function, and metabolism for expected future demands for allocation to particular traits. Such changes in response to expected demand should obviate future trade-offs among life history traits that would occur without changes in body architecture.

Architectural changes may also allow the partitioning of gathering of specific resources into different parts of the life cycle. Absence of mouthparts in some adult moths, for example, relegates all feeding to the larval stage. Changes in architecture thus may be in response to future expected allocation demands, but they also will themselves affect resource demand or intake.

B. An Insect Example: Metamorphosis

Metamorphosis in holometabolous insects includes the complete redesign of the body plan to meet adult needs. The insect changes from a "feeding machine" into a "reproductive" or "reproductive + feeding machine." Dispersal often occurs in the adult stage as well. Metamorphosis thus should result in a body well designed for reproduction and, often, dispersal.

To test the idea that allocation of larval resources during metamorphosis is adjusted to expected adult feeding intake and reproductive demand, I did a comparative study among heliconiine butterfly species (Lepidoptera: Nymphalidae) (Boggs, 1981a). The species used were *Dryas julia*, *Heliconius charitonius*, and *Heliconius cydno*. All three species will feed on the same larval host plants in the genus *Passiflora*. However, they differ significantly in adult feeding habits and amount of reproduction. Note that the hypothesis to be tested is not that animals respond to environmental variation as it is occurring, but rather that they respond to expected mean environmental conditions, which themselves vary among species in predictable ways.

Dryas julia adults feed only on nectar, obtaining carbohydrates and very few nitrogenous compounds (see Boggs, 1987). In contrast, adults of the two *Heliconius* species feed on both nectar and pollen (Gilbert, 1972). Adults collect pollen on the proboscis shortly after anther dehiscence. The pollen is held on the proboscis as a mass, and fluid is exuded by the butterfly. The mixture of pollen and fluid causes the pollen to start to germinate and release free amino acids and other compounds. The butterfly then imbibes the nutrient-rich fluid, and sloughs off the pollen, often after several hours. These species thus have a rich source of nitrogenous compounds available to the adults. Daily intake is quantifiable by rating the size of the pollen load on a scale of 0–3 each morning (Boggs, 1981a).

Females of all three species have a further adult source of nitrogen available from the spermatophore, an accessory gland product placed by the male in the female's bursa copulatrix at mating. Lepidopteran spermatophores contain proteins, carbohydrates, and lipids (see Leopold, 1976, and

Zeh and Smith, 1985, for reviews), and, in some species, contain nutrients, such as zinc (Engebretson and Mason, 1980) or sodium (Adler and Pearson, 1982; Pivnick and McNeil, 1987), or defensive compounds (Brown, 1984; Dussourd *et al.*, 1988). The spermatophore is broken down and absorbed by the female, and nutrients are used by the females for reproduction and general maintenance (Boggs and Gilbert, 1979; Boggs, 1981b). The total male contribution used by the female depends on the number of matings a female receives, and on the proportion of the spermatophore absorbed, which varies from species to species.

These heliconiine species are predicted to differ in their allocation of larval-derived resources to reproduction versus body building so as to make up the expected shortfall of resources needed by adults for reproduction. I therefore made an adult nitrogen budget for each sex and species, giving expected reproductive output and intake from pollen feeding and, for females, from spermatophores. The difference between output and intake estimated the expected minimum shortfall of nitrogen needed for reproduction, assuming all intake is allocated to reproduction rather than some other use. The nitrogen budgets were used to generate a species rank order within each sex for the proportion of larval nitrogen allocated to reproductive reserves (Fig. 3).

To test the prediction, I used Kjeldahl nitrogen analysis to obtain the amount of nitrogen in the abdomen and total body for newly emerged adults. Abdomens are primarily fat body and reproductive organs in newly emerged adult butterflies. Abdomen nitrogen thus estimated the total nitrogen available for reproduction. The ratio reserve nitrogen : total nitrogen at emergence for the heliconiine species followed the predicted rank order, and differences among all species within a sex were significant (Fig. 3). Differences in architecture seen among species (or sexes) thus are congruent with differences in expected life history and foraging traits.

Two further studies since 1981 have upheld and expanded these results. Using the butterfly *Polygonia c-album* (Nymphalidae), which has a facultative adult reproductive diapause, Karlsson and Wickman (1989) showed that allocation of larval resources during metamorphosis is adjusted to expected adult life span. More resources are allocated to the thorax as opposed to the abdomen in diapausing (longer lived) adults, suggesting that a more durable body is produced. May (1992) examined lipid allocation at metamorphosis in two butterfly species differing in foraging discrimination, and hence in net energy intake as adults. He found differences between the butterfly species in allocation of dry mass to lipids during metamorphosis as predicted on the basis of adult food intake.

C. A Plant Example: Meristems

For plants, meristem number and developmental fate controls body architecture throughout the developmental process. Variation in meristem num-

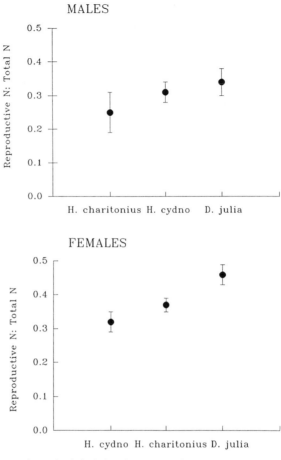

Figure 3   Mean and standard deviation for reserve nitrogen : total nitrogen at adult emergence by sex and species for three heliconiine butterflies. Species are ordered by predicted increasing ratio within each sex. Means for all species within each sex are significantly different at $P < 0.01$. Data are from Boggs (1981a).

ber and fate thus should affect possible future resource allocation to life history and foraging traits in the same way that body formation during metamorphosis does in holometabolous insects.

An excellent example of the operation of these constraints comes from work by Geber (1990) on *Polygonum arenastrum* (Polygonaceae). Apical meristems in this species are committed to reproduction or to vegetative growth early in the plant's life. Meristems committed to reproduction do not produce more apical meristems, whereas meristems destined for vegetative growth do produce further apical meristems. A commitment by a meristem to reproduction thus results in determinate growth, whereas a commitment to vegetative growth yields continuing indeterminate growth. The difference in design has a genetic component. Those individuals making an early commitment of meristems to reproduction had high early fecundity, but later growth and reproductive rates were low, due to the lack of meristems. On the other hand, individuals making an early commitment of meristems to growth had low early fecundity, but, since more meristems were available late in life, later fecundity and growth rates were high. We thus see different suites of life history traits determined by plant architecture based on meristem developmental fate.

Constraints on life history and foraging traits caused by body architecture and meristem limitation are likely to be widespread in plants. Watson (1984) was one of the first to point out that developmental constraints affect life history and foraging traits in plants. She noted that a limited meristem population produces a negative correlaton between florescence production and ramet population growth rate in *Eichhornia crassipes* (Pontederiaceae), and that similar effects are likely to be common in other species.

D. Integration and Future Directions

Developmental flexibility of body architecture allows changes in allocation that can be optimized to each age group's expected foraging and life history needs, as demonstrated in both insects and plants. This flexibility reflects techniques used by organisms to enhance allocation plasticity. However, variation still existed among individuals within a species in body architecture. In the heliconiines, this was manifested in variation within a species in abdomen nitrogen:total nitrogen; in *Polygonum,* the effect was seen in variation among individuals in early meristem fate.

The effect of such variation on individual fitness in the wild has not been fully explored. Fitness effects could be substantial, however, depending on (1) whether architectural changes are fine-tuned in response to the environment, and whether environmental variability is predictable, and/ or (2) whether variation in architecture within a group is genetically based, and subject to a fluctuating selection regime based on environmental variation. For example, the presence of a genetic component controlling early meristem fate in *Polygonum* suggests that architecture has been affected by spatial or temporal environmental variation selecting on early reproduction. Future studies are needed focusing on the effects of individual variation in architecture and the effects on allocation and fitness under natural conditions.

IV. Individual Allocation Responses to Variable Environments

A. Hypothesis

Allocation to foraging can reduce or eliminate allocation trade-offs among life history traits.

If allocation to one life history trait is increased, physiological trade-offs between that and other life history traits are often predicted. Such trade-offs are postulated, for example, between reproduction and survival, or between reproduction and growth. The underlying assumption is that the organism's resource pool is constant, so that an increase in allocation to one trait necessitates a decrease in allocation somewhere else.

Barring restrictions on resources available in the environment, there is no *a priori* reason why the resource pool must remain constant in the face of increased demand for a particular life history trait (Tuomi et al., 1983; Abrams, 1991; Boggs, 1992; Mole and Zera, 1994). If increasing allocation to foraging can produce a net increase in intake, organisms should respond to increased demand for, say, reproduction by increasing foraging. Allocation plasticity thus may act to avoid physiological trade-offs among life history traits.

The time scale of the allocation response may vary among species, depending on whether body architecture must be altered in order to increase foraging or whether behavioral changes must occur. For example, plants may need to produce more leaves or roots in order to increase foraging, and such production takes time. Animals may need to learn new foraging techniques if novel prey items are added to the diet. The effectiveness of the allocation response in the face of changes in demand should thus depend in part on the time scales of the allocation response relative to the demand. In cases where changes in demand can be anticipated, slower allocation responses may be adequate to avoid life history trade-offs.

B. An Insect Example: Male Spermatophores

Males of many insect species give females nutrients during mating, which those females then use for egg production and general maintenance (e.g., Thornhill, 1976; Boggs, 1995). In the Lepidoptera, this nutrient donation takes the form of a spermatophore, as noted above (Section III,B). Spermatophores in butterflies may represent 1.4–15.5% of a male's body weight (Rutowski *et al.*, 1983; Svärd and Wiklund, 1989) and so are a significant reproductive expenditure.

To examine the effects of increasing allocation to reproduction on life history and foraging traits, H. A. Woods *et al.* (unpublished data) manipulated the mating frequency of *H. charitonius* males and measured resulting foraging activity, life span, and spermatophore dry weight and nitrogen content. Foraging activity was determined by daily pollen load size on the proboscis, scored on a scale of 0–3 (Boggs, 1981a). Male mating treatments were (1) no mating, (2) a mating every 20 days, and (3) a mating every 10 days. The third treatment yielded a maximum number of matings equal to the maximum seen in free-flying greenhouse populations of *H. charitonius* (Boggs, 1979).

In an initial study, increasing mating frequency resulted in significant increases in pollen feeding. Effects of mating frequency on life span were inconclusive, due to a relatively small sample size. Spermatophore dry weight was not significantly affected, but nitrogen content decreased with increasing mating frequency, indicating that increased foraging was not sufficient to maintain spermatophore quality. Further analysis using another population of *H. charitonius* is in progress.

Other studies have also suggested that foraging effort may play a role in mitigating trade-offs when demand increases. For example, Partridge and Andrews (1985) note that the gradual increase in male *Drosophila melanogaster* (Diptera: Drosophilidae) daily survival rate when females are no longer available could be due to a gradual replenishment of a depleted resource pool. However, foraging was not measured in their experiment. In the sand cricket, *Gryllus firmus* (Orthoptera: Gryllidae), long-winged morphs are significantly less efficient at converting assimilated nutrients into new biomass (such as ovarian tissue) than were short-winged (nonflying) morphs, suggesting that flight is expensive to maintain and raising the possibility of a trade-off between ability to fly and fecundity, as is seen in *G. rubens* (Mole and Zera, 1993, 1994). However, when food was available *ad libitum* to *G. firmus* adults, long-winged morphs ate more than short-winged individuals, and no difference in ovarian mass was seen over the first 2 weeks of adult life. Mole and Zera (1994) note that the two morphs differ in ovarian mass after the first week of adulthood, suggesting the possibility of a trade-off that is eliminated over time and whose detection is thus dependent on the time scale of analysis. This suggests that the long-winged morphs are not immediately able to increase foraging.

C. A Plant Example: Reproductive Sexual Dimorphisms

Many plants exhibit sexual dimorphisms in total reproductive effort. The hypothesis predicts that these species should also show differences in resource intake, which may mitigate potential trade-offs involving reproduction and other life history traits.

Hebe subalpina (Scrophulariaceae) is a subdioecious evergreen shrub found in New Zealand. Plants that are male-sterile are classed as female; those that produce pollen and minimal fruit are classed as male. Female reproductive output is nearly twice that of male reproductive output (Table

Table I   Ratio of Female : Male Dry Body Mass (mg/shoot) in *Hebe subalpina* for Reproductive Structures and Vegetative Shoots[a]

	Preanthesis	Postanthesis	Total
Reproductive	0.6	4.7	1.9
Vegetative	1.4	0.8	1.1

[a] Data from Delph (1990).

I; Delph, 1990). Females produce more leaves earlier in the season than do males, so that more total resources are gathered by females over their life spans. As a result, the amount of resources allocated to growth is actually slightly larger in females than in males, and no trade-off between growth and reproduction is seen between the sexes.

Increased resource intake was also found to at least partially mitigate potential trade-offs between growth and reproduction in the grass *Agropyron repens* (Triticeae) (Reekie and Bazzaz, 1987a,b). Reproductive plants had higher net assimilation rates, due to photosynthesis of inflorescences and culms. As a result, total growth did not differ between reproductive and nonreproductive plants under most light, nutrient, and genotype treatments, although vegetative growth was reduced in reproductive plants in most cases. Plants of differing genotypes varied significantly in the amount of reproductive photosynthesis (Reekie and Bazzaz, 1987a), suggesting the possibility for selection for allocation patterns in response to reproductive demand in this species.

D. Integration and Future Directions

Allocation to foraging increased in both insects and plants in response to increased demand for reproduction, at least potentially mitigating life history trade-offs. The insect and plant examples differ, however, in the timing of the allocation response. In plants, if new tissues must be developed in order to increase foraging, time is needed to increase foraging intake. In *H. subalpina*, female plants anticipated the increased need for resources for reproduction, and they altered growth patterns prior to reproduction. *Heliconius charitonius*, on the other hand, shifted foraging effort in response to reproductive stress, rather than in anticipation of such stress.

Increases in foraging effort without architectural changes are certainly also possible in plants. In some species, leaf photosynthesis increases in response to sink strength of developing infructescences (Hartnett and Bazzaz, 1983; Tissue and Nobel, 1990; Wardlaw, 1990). However, the presence of photosynthetic reproductive organs may complicate the picture. The snow buttercup (*Ranunculus adoneus*, Ranunculaceae) has photosynthetic infructescences. Shading of infructescences results in significantly smaller achenes (smaller resource sinks), but shaded and unshaded plants show no difference in leaf photosynthetic rates (Galen *et al.*, 1993). Thus, increased resource needs for reproduction were filled by infructescence and achene photosynthesis, rather than by changes in leaf photosynthesis. Future studies on circumstances determining plasticity of allocation to foraging in the face of differing demand are needed.

V. Among-Population Responses to Variable Environments

A. Hypothesis

Predictable environmental variation across space results in genetic differentiation of allocation patterns among populations.

Predictable spatial differences in resource or time availability may form a selection gradient for resource allocation patterns within populations or species. The optimal allocation pattern to life history and foraging traits may also differ between two populations in which differing resources are limiting, such as carbon as opposed to nitrogen. In the extreme case, predictable spatial variation may result in speciation at the ends of a gradient.

Environmental gradients resulting in differences in time available for reproduction provide good opportunities to test the idea that predictable spatial variation results in differentiation of resource allocation patterns. Such gradients may be elevational or topographical, or as simple as the edge to the center of a snowbed as it melts out.

B. An Insect Example: Variation in Flight Activity Time

Colias philodice eriphyle must raise its body temperature to at least 29°C, and ideally 35°–39°C in order to fly (Watt, 1968). Body temperature is increased above ambient conditions by orienting the body perpendicular to the sun's rays. Wind speed and ambient air temperature also influence whether the butterfly can reach a body temperature suitable for flight.

Using information on the butterfly's absorptivity to sunlight, insulating characteristics, and weather conditions, Kingsolver (1983a,b) modeled flight availability time (FAT) for *C. philodice eriphyle* across an elevational gradient from 1675 to 2700 m in the Elk Mountains of Colorado. FAT decreases with increasing elevation. That is, weather conditions become less favorable for flight with increasing elevation (more clouds, more wind, and lower ambient temperatures). Although the butterflies show predictable adaptive changes with elevation in solar absorptivity and insulation, variation in weather conditions at high elevations limits the ability of the butterflies to adapt optimally to the mean—if a butterfly optimizes solar absorptivity and insulation for mean conditions at high elevation, thermal models predict that it is very likely to be killed by overheating during relatively rare periods when the sky is clear and the wind is still (Kingsolver and Watt, 1983).

Flight availability time thus varies across the elevational gradient, and resource allocation should vary in a corresponding manner. In this particular case, flight is necessary for both reproduction and foraging. Females must fly in order to find larval host plants for egg laying, and to find flowers for nectar. Springer and Boggs (1986) thus predicted that the total number of eggs laid by female *C. philodice eriphyle* should decrease with elevation, since life span and egg laying rate are constant across the elevational gradient (life span: Watt *et al.*, 1979; Tabashnik, 1980; oviposition rate: Kingsolver, 1983b; M. Stanton, personal communication, 1985). Further, since *Colias* females emerge as adults with a fixed number of oocytes (Stern and Smith, 1960), the total number of oocytes at adult emergence should decrease with increasing elevation. That is, allocation during metamorphosis to oocytes should correspond to the time available to adult females for reproduction.

Springer and Boggs (1986) studied populations of *C. philodice eriphyle* in central Colorado at 1675 and 2350 m. Populations were separated by approximately 90 km, or at least 56 times the upper estimate for Wrightian neighborhood size for populations on native host plants (Watt *et al.*, 1979). Female offspring from these populations raised under common environmental conditions showed significant differences in the mean number of oocytes, with females from the lower elevation population having more oocytes (Fig. 4). Further, the number of oocytes per female in each population closely matched predictions from Kingsolver's climate space model (Kingsolver, 1983a,b), assuming (1) a linear relationship between FAT and elevation, (2) a 5-day adult life span, (3) weather conditions allowing for maximal FAT at each elevation, (4) Stanton's mean oviposition rate, and (5) a 2 day maturation period before oviposition began.

Sample sizes for the lower elevation population were large enough to demonstrate significant within-population broad-sense heritability for oocyte number. This suggests that the trait has a genetic component, and that the differences in oocyte numbers at different elevations are at least partially genetically determined.

C. A Plant Example: Variation in Growing Season Length

There is a long history of plant studies examining ecotypic variation across environmental gradients, dating at least from the classic studies of Clausen *et al.* (1948). Many of these studies examined morphological or phenological changes across elevational gradients, but determining exact selection pressures associated with the gradients has been difficult. Water

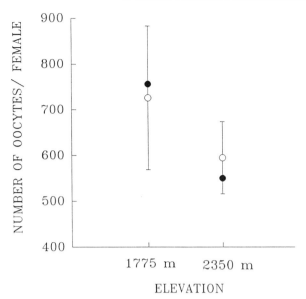

Figure 4 Predicted (●) and observed mean and standard deviation (♀) for number of oocytes at adult emergence for *Colias philodice eriphyle* as a function of elevation. Data are from Springer and Boggs (1986).

availability, soil nutrient status, temperature, wind, insolation, and growing season length all may vary across an elevational gradient. Thus, for example, Rochow (1970) grew *Thlaspi alpestre* (Cruciferae) from high and low elevation populations in the Rocky Mountains in a common garden. The two populations differed in speed of breaking and entering dormancy in response to environmental cues, with the high elevation population reacting more swiftly. This difference could be viewed as a response to shorter growing seasons at high elevations, but a quantitative understanding of the selection pressures leading to differentiation between the two populations, such as that provided by the butterfly predictive model, is missing.

Using a slightly different approach, Galen and Stanton (1993) experimentally altered growing season length for the snow buttercup, *R. adoneus*. This species is found at 3750 m in the Colorado Rocky Mountains. Location of an individual buttercup relative to the edge of a snowbed significantly affects the length of its growing season. Depending on timing of snow melt, the growing season varies between 35 and 50 days.

By manipulating snow melt schedules, Galen and Stanton (1993) altered the time available for plant growth and seed maturation. Their results indicated that both seed size and amount of plant cover were highly plastic. Seed size was directly determined by length of the growing season, with

Table II   *Ranunculus adoneus* Seed Mass under Differing Snow Bank Duration[a]

Treatment	Growing season	Seed mass (mg)	n
Control			
Snowbed edge	Long	1.2 ± 0.7	13
Snowbed center	Short	0.8 ± 0.2	8
Manipulated			
Snowbed edge delayed	Short	0.8 ± 0.2	14
Snowbed center early	Long	1.2 ± 0.2	8

[a] Data (means ± s.d.) from Galen and Stanton (1993).

longer growing seasons resulting in heavier seeds in all cases (Table II). In unmanipulated plots, percent plant cover was highest in those plots with the longer growing season. When the growing season was experimentally lengthened, percent cover increased—but an experimental decrease in growing season did not cause a concomitant decrease in percent cover. A reduced growing season thus led directly to a decrease in allocation to reproduction, but effects on plant growth could not all be accounted for by changes in available time. Galen and Stanton (1993) speculate that soil characteristics in late melt-out areas may limit growth, so that time is not the limiting factor.

D. Integration and Future Directions

In both insect and plant examples, allocation was shifted to match changes in available activity time across a gradient. However, for *Colias,* selection along the gradient has resulted in changes in allocation that have a probable genetic basis. For *Ranunculus,* allocation changes were a phenotypically plastic response; the spatial scale over which temporal variation occurs may be too small to allow genetic divergence.

Allocation changes in response to variation in resource availability across a gradient may eventually define the edge of a species' range. For *Colias,* as time for oviposition decreases, fewer oocytes are produced, with the possibility that eventually insufficient oocytes to maintain a population are made. Likewise, reductions in seed mass due to decreases in growing season might lead to difficulties in establishment of new seedlings (Galen and Stanton, 1991), defining an edge to the species' range.

The data for both insects and plants show that allocation varies within each population along an elevational gradient. This variation will interact with actually experienced environmental conditions to generate sets of individual life history parameters. Such variance in life histories among individuals can be fed into an individual-based population model, to generate predictions as to the fate of populations in the presence either of particular patterns of environmental variation or of secular trends such as those expected to be engendered by global climate change.

VI. Conclusions

Using examples from plants and butterflies (as representative insects), I have shown that (1) the ability to alter body architecture can allow optimization of allocation to meet age-specific resource demands; (2) resource pools need not be static, but may fluctuate due to allocation to foraging in response to demand for resource expenditure; and (3) spatial environmental variation may be reflected in variance in allocation, which can have a genetic basis. This exemplary set of hypotheses gives us further insight into the universality of allocation processes, and the implications for foraging strategies and population dynamics in variable environments.

Underlying many of the examples and hypotheses presented here is the further hypothesis that rates and/or efficiences of allocation will determine both the speed and ability to shift allocation in response to environmental change (Boggs, 1992). This hypothesis yields several predictions that can be tested. First, if architectural change is necessary to accomplish a particular allocation shift, then the speed of the allocation response will be a function of the organism's growth rate. Thus, selection for allocation plasticity may be manifested as selection on growth rates. Second, if allocation plasticity is accompanied by resorption of structures and reallocation of their component metabolites, or by shifting compounds from one chemical form to another, efficiency will affect the amount of resource available to be reallocated. Third, multiple resources may be limiting, or environmental variation may result in a shift among different resources as limiting. If allocation rates and efficiencies of each resource differ, the overall speed and efficiency of allocation response will be a composite of that for several resources. Examination of these areas will be critical for future integration of resource allocation, foraging, life history, and population dynamics.

Acknowledgments

I thank the organizers of the conference for inviting me to participate. F. Bazzaz, C. Malmstrom, D. Wagner, and W. Watt commented helpfully on the manuscript. This work was partially supported by the Bing Fund of the Center for Conservation Biology at Stanford.

References

Abrams, P. A. (1991). Life history and the relationship between food availability and foraging effort. *Ecology* **72**, 1242–1252.

Adler, P. H., and Pearson, D. L. (1982). Why do male butterflies visit mud puddles? *Can. J. Zool.* **60**, 322–325.

Bazzaz, F. A. (1991). Habitat selection in plants. *Am. Nat.* **137**(S), 116–130.

Bazzaz, F. A. (1996). "Plants in Changing Environments: Linking Physiological, Population and Community Ecology." Cambridge Univ. Press, London.

Bazzaz, F. A., Carlson, R. W., and Harper, J. L. (1979). Contribution to reproductive effort by photosynthesis of flowers and fruits. *Nature (London)* **279**, 554–555.

Boggs, C. L. (1979). Resource allocation and reproductive strategies in several heliconiine butterfly species. PhD dissertation. Univ. of Texas, Austin.

Boggs, C. L. (1981a). Nutritional and life-history determinants of resource allocation in holometabolous insects. *Am. Nat.* **117**, 692–709.

Boggs, C. L. (1981b). Selection pressures affecting male nutrient investment at mating in heliconiine butterflies. *Evolution* **35**, 931–940.

Boggs, C. L. (1987). Ecology of nectar and pollen feeding in Lepidoptera. *In* "Nutritional Ecology of Insects, Mites, Spiders and Related Invertebrates" (F. Slansky and J. G. Rodriguez, eds.), pp. 369–391. Wiley, New York.

Boggs, C. L. (1992). Resource allocation: Exploring connections between foraging and life history strategies. *Funct. Ecol.* **6**, 508–518.

Boggs, C. L. (1995). Male nutrient donation: Phenotypic consequences and evolutionary implications. *In* "Insect Reproduction" (S. R. Leather and J. Hardie, eds.), pp. 215–242. CRC Press, Boca Raton, Florida.

Boggs, C. L., and Gilbert, L. E. (1979). Male contribution to egg production in butterflies: Evidence for transfer of nutrients at mating. *Science* **206**, 83–84.

Brown, K. S. (1984). Adult-obtained pyrrolizidine alkaloids defend ithomiine butterflies against a spider predator. *Nature (London)* **309**, 707–709.

Calder III, W. A. (1984). "Size, Function and Life History." Harvard Univ. Press, Cambridge, Massachusetts.

Clausen, J., Keck, D. D., and Hiesey, W. M. (1948). Experimental studies on the nature of species. III. Environmental responses of climatic races of *Achillea*. *Carnegie Inst. Washington Publ.* 581. 129pp.

Delph, L. (1990). Sex differential resource allocation patterns in the subdioecious shrub *Hebe subalpina*. *Ecology* **71**, 1342–1351.

Dussourd, D. E., Ubik, K., Harris, C., Resch, J., Meinwald, J., and Eisner, T. (1988). Biparental defensive endowment of eggs with acquired plant alkaloid in the moth *Utethesia ornatrix*. *Proc. Natl. Acad. Sci. U.S.A.* **85**, 5992–5996.

Engebretson, J. A., and Mason, W. H. (1980). Transfer of ^{65}Zn at mating in *Heliothis virescens*. *Environ. Entomol.* **9**, 119–121.

Gadgil, M., and Bossert, W. H. (1970). Life historical consequences of natural selection. *Am. Nat.* **104**, 1–24.

Galen, C., and Stanton, M. L. (1991). Consequences of emergence phenology for reproductive success in *Ranunculus adoneus* (Ranunculaceae). *Am. J. Bot.* **78**, 978–988.

Galen, C., and Stanton, M. L. (1993). Short-term responses of alpine buttercups to experimental manipulaton of growing season length. *Ecology* **74**, 1052–1058.

Galen, C., Dawson, T. E., and Stanton, M. L. (1993). Carpels as leaves: Meeting the carbon cost of reproduction in an alpine buttercup. *Oecologia* **95**, 187–193.

Gatto, M., Matessi, C., and Slobodkin, L. B. (1989). Physiological profiles and demographic rates in relation to food quantity and predictability: An optimization approach. *Evol. Ecol.* **3**, 1–30.

Geber, M. A. (1990). The cost of meristem limitation in *Polygonum arenastrum:* Negative genetic correlations between fecundity and growth. *Evolution* **44**, 799–819.

Gilbert, L. E. (1972). Pollen feeding and reproductive biology of *Heliconius* butterflies. *Proc. Natl. Acad. Sci. U.S.A.* **69**, 1402–1407.

Grant, P. R., Grant, B. R., Smith, J. N. M., Abbott, I. J., and Abbott, L. K. (1976). Darwin's finches: Population variation and natural selection. *Proc. Natl. Acad. Sci. U.S.A.* **73,** 257–261.

Gurney, W. S. C., McCauley, E., Nisbet, R. M., Murdoch, W. W. (1990). The physiological ecology of *Daphnia:* A dynamic model of growth and reproduction. *Ecology* **71,** 716–732.

Hallam, T. G., Lassiter, R. R., Li, J., and Suarez, L. A. (1990). Modelling individuals employing an integrated energy response: Application to *Daphnia. Ecology* **71,** 938–954.

Hartnett, D. C., and Bazzaz, F. A. (1983). Physiological integration among intraclonal ramets in *Solidago canadensis. Ecology* **64,** 779–788.

Jacobs, M. D., and Watt, W. B. (1994). Seasonal adaptation vs physiological constraint: Photoperiod, thermoregulation and flight in *Colias* butterflies. *Funct. Ecol.* **8,** 366–376.

de Jong, G., and van Noordwijk, A. J. (1992). Acquisition and allocation of resources: Genetic (co)variances, selection, and life histories. *Am. Nat.* **139,** 749–770.

Karlsson, B., and Wickman, P. O. (1989). The cost of prolonged life: An experiment on a nymphalid butterfly. *Funct. Ecol.* **3,** 399–405.

Kingsolver, J. G. (1983a). Thermoregulation and flight in *Colias* butterflies: Elevational patterns and mechanistic limitations. *Ecology* **64,** 534–545.

Kingsolver, J. G. (1983b). Ecological significance of flight activity in *Colias* butterflies: Implications for reproductive strategy and population structure. *Ecology* **64,** 546–551.

Kingsolver, J. G., and Watt, W. B. (1983). Thermogregulatory strategies in *Colias* butterflies: Thermal stress and the limits to adaptation in temporally varying environments. *Am. Nat.* **121,** 32–55.

Kooijman, S. A. L. M., van der Hoeven, N., and van der Werf, D. C. (1989). Population consequences of a physiological model for individuals. *Funct. Ecol.* **3,** 325–336.

Leopold, R. A. (1976). The role of male accessory glands in insect reproduction. *Annu. Rev. Entomol.* **21,** 199–221.

McCauley, E., Murdoch, W. W., Nisbet, R. M., and Gurney, W. S. C. (1990). The physiological ecology of *Daphnia:* Development of a model of growth and reproduction. *Ecology* **71,** 703–715.

May, P. G. (1992). Flower selection and the dynamics of lipid reserves in two nectivorous butterflies. *Ecology* **73,** 2181–2191.

Mole, S., and Zera, A. J. (1993). Differential allocation of resources underlies the dispersal–reproduction trade-off in the wing-dimorphic cricket, *Gryllus rubens. Oecologia* **93,** 121–127.

Mole, S., and Zera, A. J. (1994). Differential resource consumption obviates a potential flight–fecundity trade-off in the sand cricket (*Gryllus firmus*). *Funct. Ecol.* **8,** 573–580.

Partridge, L. (1987). Is accelerated senescence a cost of reproduction? *Funct. Ecol.* **1,** 317–320.

Partridge, L., and Andrews, R. (1985). The effect of reproductive activity on the longevity of male *Drosophila melanogaster* is not caused by an acceleration of ageing. *J. Insect Physiol.* **31,** 393–395.

Partridge, L., and Farquhar, M. (1981). Sexual activity reduces lifespan of male fruit flies. *Nature (London)* **294,** 580–581.

Pivnick, K. A., and McNeil, J. N. (1987). Puddling in butterflies: Sodium affects reproductive success in *Thymelicus lineola. Physiol. Entomol.* **12,** 461–472.

Reekie, E. G., and Bazzaz, F. A. (1987a). Reproductive effort in plants. 1. Carbon allocation to reproduction. *Am. Nat.* **129,** 876–896.

Reekie, E. G., and Bazzaz, F. A. (1987b). Reproductive effort in plants. 3. Effect of reproduction on vegetative activity. *Am. Nat.* **129,** 907–919.

Rochow, T. F. (1970). Ecological investigations of *Thlaspi alpestre* L. along an elevational gradient in the central Rocky Mountains. *Ecology* **51,** 649–656.

Rutowski, R. L., Newton, M., and Schaefer, J. (1983). Interspecific variation in the size of the nutrient investment made by male butterflies during copulation. *Evolution* **37,** 708–713.

Springer, P., and Boggs, C. L. (1986). Resource allocation to oocytes: Heritable variation with altitude in *Colias philodice eriphyle* (Lepidoptera). *Am. Nat.* **127,** 252–256.

Stearns, S. C. (1976). Life-history tactics: A review of the ideas. *Q. Rev. Biol.* **51,** 3–47.

Stern, V. M., and Smith, R. F. (1960). Factors affecting egg production and oviposition in populations of *Colias philodice eurytheme* Boisduval (Lepidoptera: Pieridae). *Hilgardia* **29,** 411–454.

Svärd, L., and Wiklund, C. (1989). Mass and production rate of ejaculates in relation to monandry/polyandry in butterflies. *Behav. Ecol. Sociobiol.* **24,** 395–402.

Tabashnik, B. E. (1980). Population structure of butterflies. III. Pest populations of *Colias philodice eriphyle. Oecologia* **47,** 175–183.

Tatar, M., Carey, J. R., and Vaupel, J. W. (1993). Long-term costs of reproduction with and without accelerated senescence in *Callosobruchus maculatus:* Analysis of age-specific mortality. *Evolution* **48,** 1371–1376.

Thornhill, R. (1976). Sexual selection and paternal investment in insects. *Am. Nat.* **110,** 153–163.

Tissue, D. T., and Nobel, P. S. (1990). Carbon relations of flowering in a semelparous clonal desert perennial. *Ecology* **71,** 273–281.

Tuomi, J., Hakala, T., and Haukioja, E. (1983). Alternative concepts of reproductive effort, costs of reproduction and selection in life-history evolution. *Am. Zool.* **23,** 25–34.

Wardlaw, I. F. (1990). The control of carbon partitioning in plants. *New Phytol.* **116,** 341–381.

Watson, M. A. (1984). Developmental constraints: Effect on population growth and patterns of resource allocation in a clonal plant. *Am. Nat.* **123,** 411–426.

Watt, W. B. (1968). Adaptive significance of pigment polymorphisms in *Colias* butterflies. I. Variation of melanin pigment in relation to thermoregulation. *Evolution* **22,** 437–458.

Watt, W. B., Han, D., and Tabashnik, B. E. (1979). Population structure of pierid butterflies. II. A "native" population of *Colias philodice eriphyle* in Colorado. *Oecologia* **44,** 44–52.

Woods, H. A., Boggs, C. L., and Karlsson, B. Foraging and costs of reproduction: An elastic nutrient pool? Unpublished manuscript.

Zeh, D. W., and Smith, R. L. (1985). Paternal investment by terrestrial arthropods. *Am. Zool.* **25,** 785–805.

4

Biomass Allocation and Water Use under Arid Conditions

Hermann Heilmeier, Markus Erhard,
and E.-Detlef Schulze

I. Introduction

Allocation of biomass in herbaceous plants has been shown to follow predictions from optimization theory. According to economic analogs plants should allocate resources so as to maximize biomass production (Bloom *et al.*, 1985). An outcome of optimization theory is that the availability of each individual resource is capable of influencing growth of the plant, through promoting development of those organs that are involved in acquisition of whichever resources limit growth most severely. Through regulation of biomass allocation in this way, internal pool sizes of various resources, such as carbon (C) and nutrients, remain constant even when the availability of these resources happens to change (Schulze and Chapin, 1987).

This response can be interpreted in the framework of a "functional equilibrium" (Brouwer, 1963). Both empirical and mechanistic models have successfully been applied to growth and biomass partitioning under a variety of environmental conditions (Wilson, 1988). Using a mechanistic, transport-resistance approach for modeling dry matter partitioning in trees, Thornley (1991) could simulate both ontogenetic and environmentally induced changes in biomass partitioning. The resulting patterns satisfied the requirements for C and nitrogen (N) substrates of the trees in accordance with the functional equilibrium hypothesis. Dewar (1993) included

Plant Resource Allocation

93

the effects of water potential on growth and adopted the Münch flow mechanism as the basis for phloem translocation of C and N. Like a number of other models, it considers a vegetative plant to consist of two compartments only, the shoot (i.e., leaves) as the C-supplying organ and the roots as the organ acquiring nutrients and water. In most cases these models are restricted to steady-state exponential growth, where no storage or remobilization of resources occurs and all tissues are actively involved in resource uptake. Due to increased self-shading, internal concentrations of C substrates may be diminished as a function of reduced photosynthetic activity of the shoot. Consequently, the relative growth rate of the plants will decline, and biomass increment will be no longer exponential.

When plants are growing beyond the phase of exponential growth, an increasing proportion of biomass may be allocated to compartments which serve for either support and transport or storage. The development of mechanical structures which function in stabilizing the plant has been considered in growth models for vegetative herbaceous plants in a rather empirical way only (e.g., Stutzel *et al.,* 1988). In the case of biennial plants, allocation of resources both to (i) storage of carbohydrates and nitrogenous compounds versus construction of storage sites and (ii) storage tissues versus productive leaf and fine root tissues is consistent with optimization theory (Heilmeier *et al.,* 1994). However, even in longer lived herbaceous plants a great part of the biomass remains metabolically active and therefore may show a fast and rather sensitive response to changes in environmental conditions.

II. The Special Case of Woody Plants

In shrub and tree species, the formation of woody tissues in the supportive and conducting system reduces the respiratory costs per unit plant biomass compared to herbaceous plants (Schulze, 1982). However, woody tissues are not physiologically active in resource acquisition, and consequently they may be rather insensitive to perturbations in the environment. Moreover, a great part of the perennating woody tissues may reflect only past conditions for growth, and lack any physiological function at the present time (Körner, 1994). On the other hand, numerous studies have shown that the amount of conducting area, estimated as sapwood area in stems of trees, is highly correlated with the transpiring surface area of the leaves within a species (Kaufmann and Troendle, 1981; Waring *et al.,* 1982; Oren *et al.,* 1986). The leaf area/sapwood area ratio can be influenced by growing conditions and site characteristics like transpiration demand of the atmosphere (Mencuccini and Grace, 1995). Thus, growth of new woody tissues may be rather sensitive to changes in the environment.

These results can be interpreted in view of the pipe model theory considering the xylem as a collection of water conducting tubes (Shinozaki *et al.*, 1964a,b). Perttunen *et al.* (1996) adopted the pipe model hypothesis in their model of tree growth that combines tree metabolism with the spatial structure of the tree crown. They also considered loss of sapwood area by transformation into heartwood and release of functional sapwood area for further reuse by loss of transpiring area due to leaf abscission. Their model results agree fairly well with observational data on the relationship between sapwood area and foliage mass in *Pinus sylvestris*. This demonstrates the sensitivity of sapwood construction to environmental conditions and the effect of sapwood senescence on tree radius and foliage density in the crown.

According to the functional balance hypothesis, a similarly tight relationship can be expected between leaves and fine roots. For *Eucalyptus grandis* seedlings, Cromer and Jarvis (1990) could show that allocation between fine roots and leaves was dependent on the relative rates of C and N uptake. Growth of these tree seedlings followed theoretical predictions based on the concept of steady-state nutrition that the fluxes of C and nutrients are important controlling variables for biomass allocation (Ingestad and Ågren, 1991). Using a more empirical approach, Richards (1976) proposed a functional equilibrium in 1-year-old peach trees between the capacity of the root system for water uptake and the transpiratory water loss through leaves.

All these experiments on the relationship between resource availability, physiological activities, and allocation patterns have been carried out with small seedling plants under artificial conditions in the laboratory. Apart from a limited number of field studies with adult trees, in most of which only aboveground biomass is recorded (Pereira, 1990, 1994), there is no convenient experimental approach to investigate the effect of a varying supply of nutrients or water on biomass allocation for field-grown larger trees. With increasing tree size, the situation for nutrients may be different from water. Nutrients may be stored in large quantities and remobilized during a time when demand for them is high but external supply is low (Chapin *et al.*, 1990). The importance of internally stored nitrogen in furnishing growth of new leaves and twigs increases with tree age (Millard, 1993). Therefore, in older trees the tight relationship between C and N fluxes into the plant and thus the equilibrium between leaves and fine roots may be decoupled. Ontogenetic changes in allocation between root and shoot, with shoot biomass increasing relative to root biomass in older trees (Kozlowski *et al.*, 1991), might be a consequence of the relatively smaller dependence of growth on immediate external N supply with increasing tree size. Contrarily, with a few exceptions, the quantity of water that can be stored in trees is only a small proportion of the daily demand (Simonneau *et al.*, 1993). Therefore daily water losses must be nearly completely balanced by current water uptake from the soil. Thus, a tighter relationship between the transpiring leaf area and the length or surface area of fine roots can be expected under contrasting, limiting water regimes. However, the exact nature of this relationship, especially the influence of ontogenetic shifts in biomass allocation, is not knwon.

III. A Semicontrolled Experiment on Biomass Allocation and Water Use under Different Water Availabilities with *Prunus dulcis*

Almond trees [*Prunus dulcis* (Miller) D.A. Webb] were grown under arid conditions in the Negev Desert (Israel) for 1 to 4 years in pots of different volumes with a controlled water supply in order to answer the following questions:

1. How is the root–shoot ratio and the ratio between fine roots and leaf area of a woody species influenced by different amounts of available water? Do these ratios change with the age of the trees as an effect of previous water shortage? Is there an "adaptation" to a limited water supply?

2. Does the amount of biomass produced per unit of water used in evapotranspiration depend on the amount of available water?

3. Does the amount of biomass produced per unit of water used depend on the age of the trees? What is the effect of the woody biomass on water use? How do phenology of the trees and the seasonal course of transpiration demand affect water use?

A. Experimental Design and Methods

The experiment was conducted from 1984 to 1988 at the Prof. Michael Even-Ari Farm for Runoff and Desert Ecology Research at Avdat (Central Negev, Israel). Average precipitation at this site is 80 to 90 mm per year, with no rain from May to September; potential evaporation exceeds 2600 mm per year (Evenari *et al.*, 1982). Single grafted, 1-year-old almond trees (about 0.5 m high) were grown in pots of 3 m diameter and 1, 2, and 3 m depth, yielding soil volumes of 7, 14, and 21 m^3, respectively. There were nine pots per size class and three to four replicates for each age class of the trees (1, 2, 3 and 4 years). The containers were filled with homogenized local loess and watered to field capacity (27%, v/v) once per year, at the beginning of the growing season, in order to simulate natural patterns of water availability. As the amount of water given to a tree was only a function of pot size, but not of the age of the plants, trees in pots of the same volume received different amounts of water in relation to biomass depending on their age.

Each year a representative number of trees per pot size and age class was harvested including the root system. Subsamples of the soil were washed and roots separated into fine roots (diameter < 1 mm), medium roots (diameter between 1 and 5 mm), and coarse roots (diameter > 5 mm). Root length and root dry weight were taken from each fraction and used for estimating total root length and biomass of the trees. Aboveground biomass was separated into leaves and twigs/branches of different age classes. Dry weight of twigs/branches and leaves and leaf area were determined separately for each age class. During the growing period, the length increment of selected branches was recorded.

Owing to ontogenetic effects on biomass allocation, which cause ratios between two compartments to depend on plant size, data were analyzed using analysis of covariance (Packard and Boardman, 1988). Depths of the containers as the indicator for the amount of available water and age of the trees were classified as treatment effects, and the dependence of one plant compartment (e.g., shoot biomass) on them was investigated using a second one (e.g., root biomass) as the covariate. Logarithmically transformed data were used (1) to render variances homogeneous and (2) to account for possible allometric relationships between plant compartments as a consequence of differences in relative growth rates between them.

B. Effect of Tree Age and Amount of Available Water on Allocation of Biomass

Considering leaf area and length of fine roots as functional plant components, these two variables were nearly linearily related (Fig. 1). Both the amount of available water and tree age influenced this relationship interactively, although the effect was not very pronounced (Table I). The length of fine roots formed at a given biomass was independent of tree age and amount of available water (Fig. 2, Table I). The slope of the logarithmic relationship was not significantly different from 1 ($p = 0.53$). Biomass allocation within the total root system, however, was influenced by the interaction effect of tree age and amount of available water (Fig. 3, Table I). This was due to trees that were older and well supplied with water having a smaller rate of increase of fine root biomass with respect to the remainder of the root system than when trees were given a low water supply. Presumably, the larger pot volume allowed older trees to construct a more voluminous belowground system for support and storage. Alternatively, the larger structural root mass could simply be the consequence of more exploratory root growth for acquisition of water in a large soil volume over the course of several years.

Biomass partitioning between aboveground organs was also dependent on the interaction effect of tree age and amount of available water (Fig. 4, Table I). The rate of increase in leaf area with respect to wood biomass was smaller in older compared to younger trees. This effect was enhanced for trees with a low amount of available water.

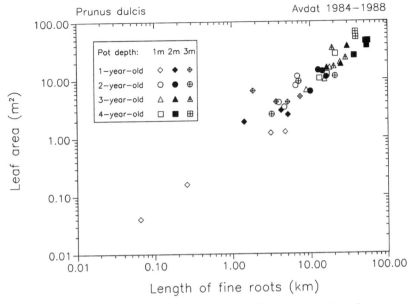

Figure 1   Relationship between leaf area and length of fine roots for almond trees grown for 1 to 4 years in pots of 1, 2, and 3 m depth.

Age of the trees and amount of available water did not affect partitioning of biomass between root and shoot, i. e., stem plus branches (Fig. 5, Table I). However, the allometric constant k (slope of the relationship) was significantly lower than 1 ($k = 0.837$, s.e. $= 0.039$; $p < 0.001$). Thus, with increasing shoot biomass, the trees allocated less biomass to roots. However, this effect was only a consequence of increasing plant size and was not due to a decreasing relative amount of available water.

C. Effect of Tree Age and Amount of Available Water on Biomass Production per Water Use

The amount of biomass produced per volume of water lost by evapotranspiration was calculated by measurement of the width of individual growth rings of 1- to 4-year-old trees during one season only. This was done in order to avoid climatic and other systematic effects that might differ between individual years. Therefore, only growth of aboveground woody biomass (stem plus branches) could be considered. Considering the whole data set, shoot biomass was linearly related to water uptake (Fig. 6). The age of the trees had a highly significant effect on this relationship (Table I). Older trees may produce up to 10 times the amount of shoot biomass per volume of water used compared to younger trees. Although one could speculate

Table I Analysis of Covariance for Relationships between Individual Plant Compartments and between Shoot Biomass Increment and Water Uptake and Water Uptake Corrected for Vapor Pressure Deficit D[a]

Effect	Available water			Tree age			$V_w \times$ age		
	df	F	P	df	F	P	df	F	P
Leaf area $= f$ (length of fine roots)	2	7.17	<0.01	3	4.14	0.02	6	3.45	0.01
Length of fine roots $= f$ (biomass of fine roots)	2	0.58	0.57	3	1.04	0.39	6	0.57	0.75
Biomass of fine roots $= f$ (biomass of medium + coarse roots)	2	1.04	0.37	3	2.61	0.07	6	2.58	0.04
Leaf area $= f$ (wood biomass in stem plus branches)	2	1.39	0.27	3	0.07	0.98	6	4.37	<0.01
Total root biomass $= f$ (biomass of shoot = stem + branches)	2	1.43	0.26	3	0.99	0.41	6	2.12	0.08
Shoot biomass increment $= f$ (water uptake)	2	3.60	0.05	3	22.2	<0.01		—	
Shoot biomass increment $= f$ (water uptake corrected for D)	2	2.36	ns	3	4.75	0.06		—	

[a] Main effects are amount of available water (V_w) as a function of lysimeter depth and age of the trees. Data were logarithmically transformed before analysis of covariance (df, degrees of freedom; F, variance ratio; P, level of significance; ns, not significant).

that a reduced relative amount of water available to older trees might have caused their higher biomass production through a physiological or morphological adaptation to water shortage, this was not the case. Older trees started and finished extension growth of branches much earlier than younger trees (see Fig. 7 for trees in a 1 m deep pot). Therefore, their main growth period was during a time of relatively low potential evapotranspiration in spring, compared to very dry conditions in summer. Leaf area production per water used was significantly negatively correlated with the length of the growth period (Fig. 8A; $r = -0.53$, $p = 0.01$, $n = 22$) and with average vapor pressure deficit of the air during the growth period (Fig. 8B; $r = -0.55$, $p < 0.01$, $n = 22$). Thus, when biomass production per water use was corrected for differences in potential evapotranspiration, neither the age of trees nor the amount of available water had a significant effect (Table I). Therefore, an increase in the amount of biomass produced per water used was a consequence of different rates of developmental processes. These were due to the positive effect of perennating woody biomass, which allowed a high rate of growth early in the season when potential evapotranspiration was still low.

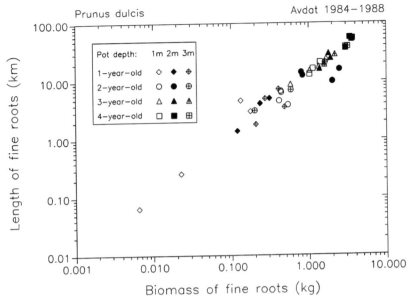

Figure 2 Relationship between length and biomass of fine roots for almond trees grown for 1 to 4 years in pots of 1, 2, and 3 m depth.

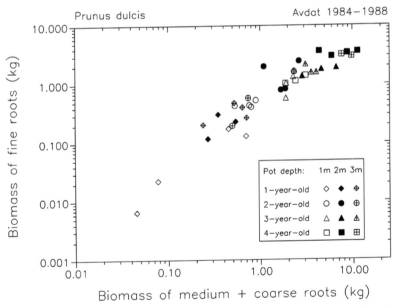

Figure 3 Relationship between biomass of fine roots (diameter < 1 mm) and of medium plus coarse roots (diameter > 1 mm) for almond trees grown for 1 to 4 years in pots of 1, 2, and 3 m depth.

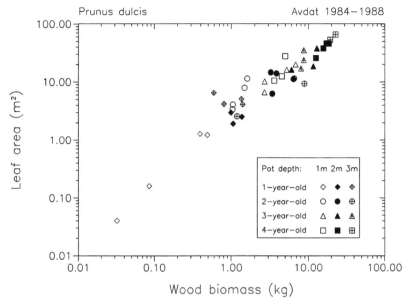

Figure 4   Relationship between leaf area and biomass of wood in stem and branches for almond trees grown for 1 to 4 years in pots of 1, 2, and 3 m depth.

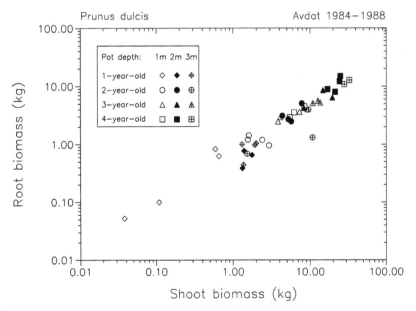

Figure 5   Relationship between total root biomass and shoot biomass (stem and branches) for almond trees grown for 1 to 4 years in pots of 1, 2, and 3 m depth.

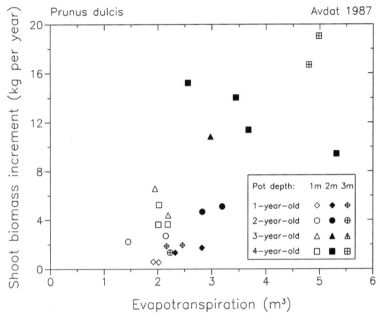

Figure 6   Relationship between shoot biomass increment (kg per year) and water uptake for almond trees planted in four successive years in pots of 1, 2, and 3 m depth and harvested in 1987.

D. What Is the Effect of the Rate of Developmental Processes on the Amount of Biomass Produced per Water Used and Biomass Allocation?

Having identified phenology of trees as the dominant factor influencing the relationship between biomass production and water use, one might speculate about the effect of accelerated leaf growth rates for a tree with a given water supply. Growth of leaves early in the season, when potential evapotranspiration is still low, should allow more leaf and wood biomass to be produced per amount of water taken up as a consequence of small transpiratory losses. That means that for a given amount of water supplied to the trees, a higher rate of leaf growth should elevate the increase of leaf area and biomass in any given season. On the other hand, a higher rate of leaf area increase could cause total tree transpiration to rise on any given day because the increase in transpiring surface might compensate for any savings of water due to low evaporative demands. Increased transpiration requires a higher water uptake capacity, i.e., more fine roots, which has a negative effect on whole plant C balance. This trade-off was modeled for a given tree with a limited amount of available water under different rates of simulated leaf area growth.

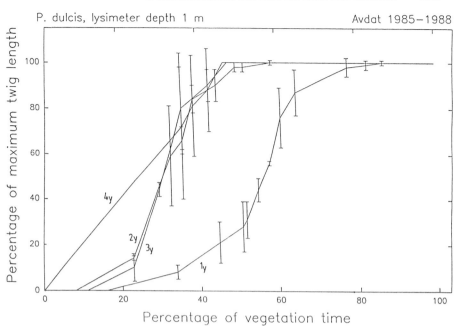

Figure 7   Percentage of maximum twig length per age class of trees (1y, 1-year-old trees; 2y, 2-year-old trees; 3y, 3-year-old trees; 4y, 4-year-old trees) as a function of percentage vegetation time (100% = vegetation time of all age classes pooled, end of February to beginning of September). Error bars indicate standard errors.

The relative expansion rate of leaf area as dependent on soil water potential was allowed to vary from actual measured values to twice these values (Fig. 9A). According to the mean vapor pressure deficit D during the growth period, which decreased with increasing rates of leaf area growth, different amounts of leaf area could be produced per volume of water taken up (Fig. 9B). Cumulative water uptake determined plant water status (expressed as predawn leaf water potential, Fig. 9C), which in turn affected total daily carbon gain. Due to stomatal closure, total daily C gain was reduced during days with a high vapor pressure deficit compared to low D (Fig. 9D). Water uptake, on the other hand, required a minimum amount of root length, which was estimated by a steady-state, single root model for radial water flow according to Gardner (1960) using the following equations:

$$\psi_s - \psi_r = J_w \times R_{sr};$$

$$R_{sr} = -\frac{\ln(L_v \times \pi \times r_r^2)}{4 \times \pi \times k(\theta) \times L_v};$$

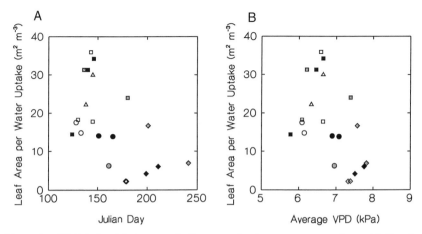

Figure 8   (A) Relationship between leaf area production per water uptake and Julian day number for the end of the growth period (different symbols are for different pot depth ×tree age combinations; see Figs. 1–6). (B) Relationship between leaf area production per water uptake and average vapor pressure deficit of the air (VPD) during the growth period (symbols as in Figs. 1–6).

where ψ_s and ψ_r are soil and root water potential respectively, J_w is the water flux to a single root, R_{sr} is the hydraulic resistance between soil and root, $k(\theta)$ is the unsaturated soil hydraulic conductivity, r_r is root radius, and L_v is root length density (length of fine roots per soil volume). An effective root length density was calculated using the root contact concept of Herkelrath *et al.* (1977), in which varying degrees of contact between root and soil as a function of soil water content are considered. Length of fine roots was converted to fine root biomass using specific root length. Total root biomass was calculated using the allometric relationships given above (Fig. 3). Construction and maintenance costs were estimated using data from the literature. These costs for root growth were compared with total cumulative C gain in order to yield C available for growth of heterotro-

Figure 9   Schematic diagram of the gas exchange–water relations–growth model for the carbon gain submodel. (A) Dependence of relative leaf area expansion rate ($m^2 m^{-2} day^{-1}$) on soil water potential [$\log(-0.1 \times MPa)$]. (B) Relationship between cumulative amount of water evapotranspired and cumulative leaf area growth. (C) Dependence of predawn leaf water potential on cumulative amount of water evapotranspired. (D) Dependence of daily carbon gain on predawn leaf water potential. Solid curves are fitted to experimental data. Dashed curves are for the maximum range of simulated leaf area expansion rate (A) or indicate the effect of lowered vapor pressure deficit due to accelerated leaf growth (B and D).

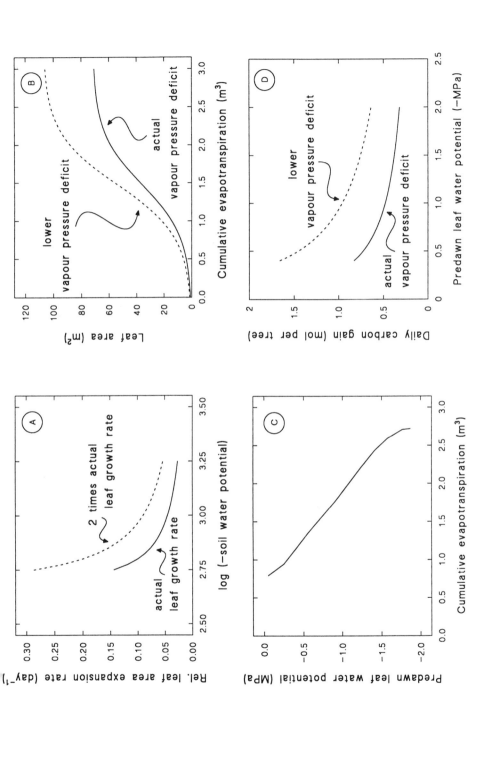

Table II   Carbon Gain and Carbon Costs Related to Different Growth Periods[a]

Length of growth period (days)	Evaporative demand (kPa × days)	Leaf area (m²)	C gain (kg)	C costs for fine roots (kg)	C costs for total roots (kg)	C available for growth of wood (kg)	C in wood (kg)
225	177	66.6	12.2	1.97	6.1	6.1	6.2
190	128	73.3	12.2	2.25	11.1	1.1	6.8
177	108	80.3	12.3	2.32	13.8	−1.5	7.4
161	82.8	94.1	12.4	2.15	12.4	−0.2	8.6
144	57.8	120	12.5	3.22	102	−90	10.8

[a] The length of the growth period was calculated as a function of different rates of leaf area growth relative to the rate actually measured for the trees (acceleration factors 1, 1.125, 1.25, 1.5, and 2 for rows 1 to 5), assuming a constant amount of water given to the trees. Evaporative demand was estimated by integration of a sinoidal function fitted to measured values of Class A pan evaporation. Leaf area was modeled according to Fig. 9. For calculation of C gain and C costs, see text. Carbon in wood was estimated by elemental analysis of aboveground woody biomass for the tree with a growth period of 225 days and modeled for shorter growth periods by allometric relationships between leaf area and aboveground wood biomass (see Fig. 4).

phic aboveground structures (all leaf growth was assumed to be supported by internal storage of C). The biomass of stem and branches constructed for a given rate of leaf area growth was estimated from allometric relationships between leaf area and wood biomass (Fig. 4).

Model results are shown in Table II. Data in the first row of Table II describe the actually observed length of the growth period, and data in rows 2 to 5 are for relative leaf area expansion rates increased by a factor of 1.125, 1.25, 1.5, and 2.0, respectively. For a given amount of water supplied to the trees, a higher rate of leaf area expansion, yielding in a higher total leaf area and therefore higher transpiratory water loss, reduces the length of the growth period. Consequently, total evaporative demand ($\int D\, dt$) is decreased by nearly 70% for the shortest growing period compared to the actually observed length of the growth period. Total leaf area produced with the same amount of water (3.2 m³) is increased by 80% for the highest rate of leaf area growth. However, total C gain is nearly unaffected by the increase in leaf area. This is a consequence of the high rate of transpiration with increasing leaf area, which reduces predawn leaf water potential faster than for plants with a lower rate of leaf area expansion. Thus, total daily C gain is drastically diminished especially during a time when total leaf area is high. High transpiration rates in plants with high rates of leaf growth require an extensive fine root system, which increases C costs for fine root growth by 65% for fastest leaf area growth. Due to the allometric relationship between biomass of fine roots and biomass of medium plus coarse roots, C costs for the total root system are increased nearly 20-fold for the highest rate of leaf growth, which causes total plant C balance to become unrealistically negative. Even for an acceleration factor of 1.125 to 1.5, however, total C gain would approximately cover C costs only for root growth if the allometric relationship within the root system holds for different rates of leaf area expansion. Only in the case of the actually observed rate of leaf growth did C in wood biomass nearly equal the amount of C available for aboveground growth after root growth was accounted for. Thus, this simulation suggests that accelerated leaf growth would cause C costs to the plant for enhanced root growth that cannot be covered even under reduced transpiration demand. Therefore, the almond trees studied in this experiment seem to grow at their maximum possible rate, if a positive C balance is considered to represent the major factor determining growth rate.

IV. Conclusions

The aim of the experiment with almond trees was to study principles of biomass allocation of a woody species over several growing seasons under a reduced water supply which simulated natural rainfall patterns in the study area. For this reason, trees were given water only once per year at the beginning of the growing season. The experimental design differed from a number of studies in the literature, in which a group of trees is constantly given a stated proportion of the amount of water received by a control group (see, e.g., Steinberg *et al.*, 1990, for peach trees). In another type of experiment water is withheld for a group of trees, which is compared to a control group being fully watered (see, e.g., Khalil and Grace, 1992, for sycamore). However, controlled atmospheric conditions during the course of these experiments caused parallel changes in transpiration demand for both well-watered and drought-stressed plants. Enhanced root growth relative to leaf growth in droughted plants during these experiments can be interpreted to support the establishment of a functional balance between shoot and root at a reduced water supply when water uptake per unit of root length decreases. A reduction in leaf area per unit of root length was accompanied by decreases in stomatal opening, thus reducing transpiratory water loss even further. As a result, leaf water potential of droughted plants may not differ from control plants (Khalil and Grace, 1993). Thus, both physiological and morphological responses of a plant may be integrated to keep leaf water status at a level that will not impair metabolic processes like photosynthetic C gain (Schulze *et al.*, 1983; Givnish, 1986).

However, steady-state conditions in these experiments do not consider changes in environmental conditions over time, as was realized in the field experiment with almond trees described here. Trees of various age and water availabilities suffered different rates of soil drying at different evaporative demands of the atmosphere. Therefore, different groups of plants would have experienced contrasting soil water conditions at the same time, whereas water availability would have been similar at various points in time with contrasting atmospheric conditions. These changing combinations of environmental conditions require a high flexibility of plant responses at various levels (Mooney and Winner, 1991). Stomatal response to soil water status has been shown to be mediated by abscisic acid (ABA) transported from roots to leaves in the transpiration stream of the almond trees in this experiment (Wartinger *et al.*, 1990). However, Munns and Sharp (1993) found no conclusive evidence that root-sourced ABA regulates leaf expansion in dry soil. Therefore, control of the equilibrium between leaf and root growth must occur on a higher, more integrative level. Maintenance of plant internal circulation via phloem and xylem has been suggested as a possible mechanism (Schulze, 1994).

This mechanism of maintenance of internal circulation may be responsible for the long-term homeostasis in growth between roots and leaves over a wide range and combination of environmental conditions experienced by the almond trees. One might speculate that the trees adjust their growth to the amount of water available, and that—apart from osmotic adjustment—a certain amount of root growth is required to prevent shoot water status from declining to values detrimental to leaf metabolism. Plants with a relatively high amount of water available may extend their growth phase into a period when evaporative demand of the atmosphere is high, thus requiring a high water uptake capacity to maintain the high transpiration rates. In plants with a poor water supply, growth is restricted within the period of low potential evapotranspiration, but the low availability of soil water requires extensive growth of fine roots. Thus, maximization of C gain under given environmental conditions and finally growth may be achieved by keeping the growth rate and allocation pattern in equilibrium. Accelerated leaf growth negatively affects whole plant C balance by requiring high root growth as a consequence of exhausting water supplies. Reduced growth rates will have the same effect by diminishing C gain during high evaporative demand. The model results suggest that the almond trees show a growth rate which might allow maximum growth for a given water supply. This pattern cannot be explained by a "functional equilibrium" between leaves and roots alone as pointed out by Hilbert (1990). Rather, allocational patterns in combination with physiological responses maintain optimum leaf functioning for a given set of environmental conditions. As these may change from day to day, models on optimization of growth and timing of resource allocation (Lerdau, 1992) have to include seasonality in the environment and phenology of plants.

Acknowledgments

This work was supported by Deutsche Forschungsgemeinschaft (SFB 137). We thank people on the Avdat Farm and the Jacob-Blaustein-Institute for Desert Research (Sede Boker) for generous help during the experiments.

References

Bloom, A. J., Chapin III, F. S., and Mooney, H. A. (1985). Resource limitation in plants—An economic analogy. *Annu. Rev. Ecol. Syst.* **16**, 363–392.

Brouwer, R. (1963). Some aspects of the equilibrium between overground and underground plant parts. *Jaarb. Inst. Biol. Scheik. Onderz.*, 31–39.

Chapin III, F. S., Schulze, E.-D., and Mooney, H. A. (1990). The ecology and economics of storage in plants. *Annu. Rev. Ecol. Syst.* **21**, 423–447.

Cromer, R. N., and Jarvis, P. G. (1990). Growth and biomass partitioning in *Eucalyptus grandis* seedlings in response to nitrogen supply. *Aust. J. Plant Physiol.* **17**, 503–515.

Dewar, R. C. (1993). A root–shoot partitioning model based on carbon–nitrogen–water interactions and Münch phloem flow. *Funct. Ecol.* **7**, 356–368.

Evenari, M., Shanan, L., and Tadmor, N. (1982). "The Negev: The Challenge of a Desert." 2nd Ed. Harvard Univ. Press, Cambridge, Massachusetts.

Gardner, W. R. (1960). Dynamic aspects of water availability to plants. *Soil Sci.* **89**, 63–73.

Givnish, T.J. (1986). Optimal stomatal conductance, allocation of energy between leaves and roots, and the marginal cost of transpiration. *In* "On the Economy of Plant Form and Function" (T. J. Givnish, ed.), pp. 171–213. Cambridge Univ. Press, Cambridge.

Heilmeier, H., Freund, M., Steinlein, T., Schulze, E.-D., and Monson, R. K. (1994). The influence of nitrogen availability on carbon and nitrogen storage in the biennial *Cirsium vulgare* (Savi) Ten. I. Storage capacity in relation to resource acquisition, allocation and recycling. *Plant Cell Environ.* **17**, 1125–1131.

Herkelrath, W. N., Miller, E. E., and Gardner, W. R. (1977). Water uptake by plants. II. The root contact model. *Soil Sci. Soc. Am. J.* **41**, 1039–1043.

Hilbert, D. W. (1990). Optimization of plant root:shoot ratios and internal nitrogen concentration. *Ann. Bot.* **66**, 91–99.

Ingestad, T., and Ågren, G. I. (1991). The influence of plant nutrition on biomass allocation. *Ecol. Appl.* **1**, 168–174.

Kaufmann, M. R., and Troendle, C. A. (1981). The relationship of leaf area and foliage biomass to sapwood conducting area in four subalpine forest tree species. *For. Sci.* **27**, 477–482

Khalil, A. A. M., and Grace, J. (1992). Acclimation to drought in *Acer pseudoplatanus* L. (sycamore) seedlings. *J. Exp. Bot.* **43**, 1591–1602.

Khalil, A. A. M., and Grace, J. (1993). Does xylem sap ABA control the stomatal behaviour of water-stressed sycamore (*Acer pseudoplatanus* L.) seedlings? *J. Exp. Bot.* **44**, 1127–1134.

Körner, Ch. (1994). Biomass fractionation in plants: A reconsideration of definitions based on plant functions. *In* "A Whole Plant Perspective on Carbon–Nitrogen Interactions" (J. Roy and E. Garnier, eds.), pp. 173–185. SPB Academic Publ., The Hague.

Kozlowski, T. T., Kramer, P. J., and Pallardy, S. G. (1991). "The Physiological Ecology of Woody Plants." Academic Press, San Diego.

Lerdau, M. (1992). Future discounts and resource allocation in plants. *Funct. Ecol.* **6**, 371–375.

Mencuccini, M., and Grace, J. (1995). Climate influences the leaf area/sapwood area ratio in Scots pine. *Tree Physiol.* **15**, 1–10.

Millard, P. (1993). A review of internal cycling of nitrogen within trees in relation to soil fertility. In "Optimization of Plant Nutrition" (M. A. C. Fragoso and M. L. van Beusichem, eds.), pp. 623–628. Kluwer Academic Publ., Dordrecht, The Netherlands.

Mooney, H. A., and Winner, W. E. (1991). Partitioning response of plants to stress. In "Response of Plants to Multiple Stresses" (H.A. Mooney, W.E. Winner, and E.J. Pell, eds.), pp. 129–141. Academic Press, San Diego.

Munns, R., and Sharp, R. E. (1993). Involvement of abscisic acid in controlling plant growth in soils of low water potential. *Aust. J. Plant Physiol.* **20**, 425–437.

Oren, R., Werk, K. S., and Schulze, E.-D. (1986). Relationships between foliage and conducting xylem in *Picea abies* (L.) Karst. *Trees* **1**, 61–69.

Packard, G. C., and Boardman, T. J. (1988). The misuse of ratios, indices, and percentages in ecophysiological research. *Physiol. Zool.* **61**, 1–9.

Pereira, J. S. (1990). Whole plant regulation and productivity in forest trees. In "Importance of Root to Shoot Communication in the Responses to Environmental Stress" (W. J. Davies and B. Jeffcoat, eds.), Monogr. 21, pp. 237–250. British Society for Plant Growth Regulation, Bristol.

Pereira, J. S. (1994). Gas exchange and growth. In "Ecophysiology of Photosynthesis" (E.-D. Schulze and M. M. Caldwell, eds.), Ecological Studies 100, pp. 147–181. Springer-Verlag, Berlin.

Perttunen, J., Sievänen, R., Nikinmaa, E., Salminen, H., Saarenmaa, H., and Väkevä, J. (1996). LIGNUM: A tree model based on simple structural units. *Ann. Bot.* **77**, 87–98

Richards, D. (1976). Root–shoot interactions: A functional equilibrium for water uptake in peach [*Prunus persica* (L.) Batsch.]. *Ann. Bot.* **41**, 279–281.

Schulze, E.-D. (1982). Plant life forms and their carbon, water and nutrient relations. In "Encyclopedia of Plant Physiology, New Series, Volume 12B: Physiological Plant Ecology II, Water Relations and Carbon Assimilation" (O. L. Lange, P. S. Nobel, C. B. Osmond, and H. Ziegler, eds.), pp. 615–676. Springer-Verlag, Berlin.

Schulze, E.-D. (1994). The regulation of plant transpiration: Interactions of feedforward, feedback, and futile cycles. In "Flux Control in Biological Systems" (E.-D. Schulze, ed.), pp. 203–235. Academic Press, San Diego.

Schulze, E.-D., and Chapin III, F. S. (1987). Plant specialization to environments of different resource availabilities. In "Potentials and Limitations of Ecosystem Analysis" (E.-D. Schulze and H. Zwölfer, eds.), pp. 120–148. Springer-Verlag, Berlin.

Schulze, E.-D., Schilling, K., and Nagarajah, S. (1983). Carbohydrate partitioning in relation to whole plant production and water use of *Vigna unguiculata* (L.) Walp. *Oecologia* **58**, 169–177.

Shinozaki, K., Yoda, K., Hozumi, K., and Kira, T. (1964a). A quantitative analysis of plant form: The pipe model theory. I. Basic analyses. *Jpn. J. Ecol.* **14**, 97–105.

Shinozaki, K., Yoda, K., Hozumi, K., and Kira, T. (1964b). A quantitative analysis of plant form: The pipe model theory. II. Further evidence of the theory and its application to forest ecology. *Jpn. J. Ecol.* **14**, 133–139.

Simonneau, T., Habib, R., Goutouly, J.-P., and Huguet, J.-G. (1993). Diurnal changes in stem diameter depend upon variations in water content: Direct evidence in peach trees. *J. Exp. Bot.* **44**, 615–621.

Steinberg, S. L., Miller, J. C., Jr., and McFarland, M.J. (1990). Dry matter partitioning and vegetative growth of young peach trees under water stress. *Aust. J. Plant Physiol.* **17**, 23–36.

Stutzel, H., Charles-Edwards, D. A., and Beech, D. F. (1988). A model of the partitioning of new aboveground dry matter. *Ann. Bot.* **61**, 481–487.

Thornley, J. H. M. (1991). A transport–resistance model of forest growth and partitioning. *Ann. Bot.* **68**, 211–226

Waring, R. H., Schroeder, P. E., and Oren, R. (1982) Application of the pipe model theory to predict canopy leaf area. *Can. J. For. Res.* **12**, 556–560.

Wartinger, A., Heilmeier, H., Hartung, W., and Schulze, E.-D. (1990). Daily and seasonal courses of leaf conductance and abscisic acid in the xylem sap of almond trees [*Prunus dulcis* (Miller) D.A. Webb] under arid conditions. *New Phytol.* **116,** 581–587.

Wilson, J. B. (1988). A review of evidence on the control of shoot:root ratio, in relation to models. *Ann. Bot.* **61,** 433–449.

5

Organ Preformation, Development, and Resource Allocation in Perennials

Monica A. Geber, Maxine A. Watson,
and Hans de Kroon

I. Introduction

In plants, many life history traits, including the schedules of growth and reproduction, are the outcome of an interaction between a program of development and of resource uptake and use (Watson, 1984; Geber, 1990; Diggle, 1994; Geber et al., 1997). Unlike most animals, where postembryonic development consists of the elaboration of a fixed set of organs, plants continue to add and shed organs throughout life. The number and type of organs that are added depend on developmental events concerning the production and fate of meristems. Vegetative growth results from the elaboration of new or quiescent meristems into leaves, shoots, and roots, whereas flowering depends on the conversion of meristems to a sexual fate. The schedule of flowering and growth is dependent on resources because meristem production and fate are affected by a plant's resource status, or by correlates of resource status, such as plant hormonal balance (Takeda et al., 1980; Marino and Greene, 1981; Goldschmidt and Colomb, 1982; Diggle, 1994; Geber et al., 1997), and because resources are required to construct the organs arising from meristems. The dependence of development on resources explains why life history characteristics in plants are more often tied to the size (i.e., resources) than to the age of individuals (Lacey, 1986).

Just as resources affect plant developmental decisions, these same decisions affect plant resource relations. The open-ended nature of plant development means that plants, unlike many animals, do not have a fixed set of "mouths" through which to obtain resources, but change in their ability to acquire resources as they add and shed organs (Watson, 1984; Watson and Casper, 1984; Harper, 1989; Lerdau, 1992). Developmental commitments to leaves and roots increase the number of "mouths" for future resource gains, whereas commitments to flowers are likely to lead to increased demands on resources. Indeed, plants have been analogized to business firms (Bloom *et al.*, 1985) in which the ultimate currency of profit is lifetime fitness. Commitment of meristems and resources to vegetative structures is equivalent to a capital investment whose profit is realized in terms of increased future biomass and reproductive potential, albeit at the expense of immediate fitness gains through reproduction. Commitment to reproduction, on the other hand, can yield immediate fitness returns, but, by virtue of reducing meristem availability and resources for vegetative growth (Watson, 1984; Geber, 1990), may lower future reproductive potential.

Dating back to Fisher (1930) and Cole (1954), evolutionary ecologists have sought to determine the optimal schedules of growth, reproduction, and survival for organisms. In a world of limiting resources, this problem has been restated as one involving the allocation of scarce resources to competing life history functions in such a way as to maximize lifetime fitness. In the past 20 years, biologists have adopted the tools of optimal control theory and dynamic programming to address questions of optimal resource allocation in plants (see Kozlowski, 1992, and Perrin and Sibly, 1993, for reviews). These analytical and computational methods are appropriate because, as already noted, allocation "decisions" made in the present time have impacts both on the plant's immediate resource relations and fitness returns and on future resource status and fitness gains (Kozlowski, 1992). These and similar models have been applied with considerable success to a wide variety of allocation problems, including the timing of reproduction in annuals (Cohen, 1971; Vincent and Pulliam, 1980; King and Roughgarden, 1982a,b; Schaffer *et al.*, 1982; Kozlowski and Ziolko, 1988; Amir and Cohen, 1990; Iwasa, 1991), the schedules of growth, storage, and reproduction in annuals and perennials (Chiariello and Roughgarden, 1984; Kozlowski and Uchmanski, 1987; Kozlowski and Wiegert, 1987; Pugliese, 1987, 1988; Iwasa and Cohen, 1989; Pugliese and Kozlowski, 1990; Kozlowski, 1992), allocation to herbivore defense (Lerdau, 1992), resource partitioning between roots and shoots (Iwasa and Roughgarden, 1984), and physiological integration between ramets in clonal plants (Caraco and Kelly, 1991).

Implicit in the structure of all of the models is the assumption that the allocation of resources to a particular life history function (e.g., growth, reproduction) is synonymous with the developmental commitment of meristems to these same functions. Because of this assumption, developmental decisions need not be treated separately from allocation decisions, and the models, which are already complex, can be simplified. This assumption, however, may not be valid in plants where the developmental commitment of meristems to organs does not coincide with the allocation of resources to those organs. Thus, in many perennial plants of seasonal environments, the developmental commitment of meristems to vegetative or reproductive organs takes place months or years before the organs are elaborated, and may precede by a considerable amount of time the investment of resources to build the organs.

The syndrome of organ preformation raises a number of questions, including (1) how common is it and what is its taxonomic and ecological distribution, (2) what is the relationship between the timing of development and the timing of resource acquisition and use, and (3) what are the life history and demographic consequences of preformation? We address the first two questions by examining the prevalence and phenological associations of floral preformation in temperate forest herbs, as documented by Randall (1952). To address the third question, we use demographic simulations to examine the consequences of organ preformation in mayapple, *Podophyllum peltatum* (Berberidaceae), a forest herb from eastern North America, for which there is evidence of variation in the timing of organ preformation.

II. Organ Preformation in Plants

There are few community-wide studies of organ preformation, as the literature on this topic consists mainly of developmental studies on single species. It is difficult therefore to establish with much confidence the taxonomic and ecological distribution of this syndrome. A brief survey of the literature does indicate that, in many gymnosperms and perennial angiosperms of all life forms, shoots, cones, or flowers are at least partially preformed one or more years in advance of their expansion (Foerste, 1891; Moore, 1909; Raunkiaer, 1934; Sørensen, 1941; Randall, 1952; Sacher, 1954; Parke, 1959; Critchfield, 1960; Dafni *et al.*, 1981; Fulford, 1966; Clausen and Kozlowski, 1970; Bliss, 1971; Gill, 1971; Sattler, 1973; Owens and Molder, 1973a,b, 1979; Marks, 1975; Bierzychudek, 1982; Inouye, 1986; Nobel, 1987; Yoshie and Yoshida, 1989; Merrill, 1990; Steeves and Steeves, 1990; Newell, 1991; Westwood, 1993). Preformed buds may occur on corms, bulbs, stems, or roots, depending on the species. The literature also suggests that preformation is especially common in seasonal environments, both seasonally cold (Sørenson, 1941; Randall, 1952; Bliss, 1971; Yoshie and Yoshida, 1989) and seasonally dry (Dafni *et al.*, 1981), although this observation may result in part from researcher bias in the taxa studied.

A. Floral Preformation in Forest Herbs

In a remarkable study of forest herbs, Randall (1952) recorded data on the morphology, phenology, life history, and the distribution and abundance of all herb species encountered in 116 upland forest stands in northern Wisconsin (Brown and Curtis, 1952) and in 96 upland stands in southern Wisconsin (Curtis and McIntosh, 1951). The stands ranged from early successional to climax forest stands. Randall noted for each species whether or not preformed flower buds were present on perennating organs in the fall preceding the year of flowering. The types of perennating organs in his sample of herbs included bulbs, corms, stem buds (both above and below ground), and root and crown buds. Randall also noted (1) the season of flowering (spring, summer, fall), (2) the season of fruiting (spring, summer, fall), and (3) the season of photosynthesis as judged by the timing of leaf emergence and leaf duration (evergreen, spring, summer). Randall's phenological categories are no doubt imprecise in failing to more narrowly define the periods of maximal flowering, fruiting, and especially carbon gain of species. Nevertheless, it is possible to use the data to address questions concerning the prevalence and taxonomic distribution of floral preformation in forest herbs. One can also ask whether the advanced commitment to flowering is related to the timing of flowering and fruiting in the following season. Likewise, one can ask whether floral preformation is associated with the season of carbon gain, because carbon is a resource that is most likely to be limiting in forest understories.

Floral preformation is extremely common in the perennial herbs from Randall's survey, being present in a full 58% of angiosperm species from northern forests ($N = 98$ species) and 49% of the angiosperm species from southern forests ($N = 102$) (Table I). The difference in preformation frequency between the two forest herb communities is due to a change in species composition and not to intraspecific variation in the occurrence of floral preformation. Although not statistically significant, the shift in frequency between northern and southern communities is intriguing because it suggests that preformation may be more common at higher latitudes where growing seasons are shorter (Bliss, 1971; Yoshie and Yoshida, 1989; see below).

Floral preformation is a phylogenetically conservative trait that appears to vary mostly at the generic level. Thus, in the total sample of 141 angiosperm species from northern and southern forests, only 10% of the variation in the presence/absence of floral preformation is attributable to differences among species within a genus, and 53% of the variation is accounted for by differences among genera within families, with the remaining variation

Table I Frequency of Floral Preformation and of Phenological Classes of Flowering, Fruiting, and Photosynthetic Activity in Forest Herbs of Northern and Southern Wisconsin[a]

	Northern Wisconsin	Southern Wisconsin
Floral preformation		
Yes	58	49
No	41	51
Flowering season		
Spring	37	35
Summer	60	59
Fall	3	6
Fruiting season		
Spring	11	11
Summer	58	50
Fall	31	38
Photosynthetic season		
Spring	5	7
Summer	68	77
Evergreen	27	15

[a] Data from Randall (1952).

(37%) occurring among families (Table II). Other phenological traits related to reproduction and carbon gain are also phylogenetically conservative but the amount of variation accounted for by family and genus differs among traits (Table II). These results are consistent with findings from other studies that indicate a strong phylogenetic component to the phenology of flowering (Kochmer and Handel, 1986) and of photosynthetic activity (Lechowicz, 1984; Givnish, 1987).

Floral preformation is most strongly associated with the season of flowering ($\chi^2 = 65.7$, $df = 2$, $P < 0.0000$; Fig. 1), with preformation being most common in spring-flowering species and least common in fall-flowering

Table II Percentage of Variation in Developmental and Phenological Traits Attributable to Differences among Plant Families, Genera within Families, and Species within Genera[a]

Source of variation	Flower preformation	Flowering season	Fruiting season	Photosynthetic season
Families	37	25	18	45
Genera within families	53	49	70	19
Species within genera	10	26	12	36

[a] Based on a sample of 141 angiosperm species of perennial forest herbs from Wisconsin. From Randall (1952).

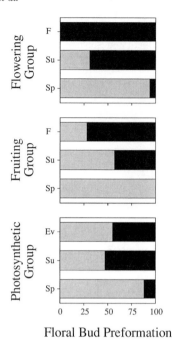

Floral Bud Preformation

Figure 1   Association between floral preformation and season of flowering, fruiting, and photosynthesis in forest herbs. Data come from a survey of herbs in Wisconsin upland forests (Randall, 1952; see text). In each graph, the percentage of species with (gray bars) and without (black bars) flower preformation is shown for each phenological group of species depicted on the *y* axis. Sp, Spring; Su, summer; F, fall; Ev, evergreen.

species. The association of floral preformation with spring flowering suggests that its advantage lies in permitting the rapid expansion of flowers early in the growing season when temperatures are generally low. The benefit of rapid expansion would also apply to preformed vegetative structures and would be especially important in forest herbs that make a substantial portion of their carbon gain in early spring before canopy closure (Yoshie and Yoshida, 1987, 1989). However, Randall did not record whether or not leaves were also preformed. Rapid organ expansion would also be advantageous in seasonally dry environments where the favorable period for growth and reproduction is limited by drought (Raunkiaer, 1934).

Because plants must flower before they set fruit, the phenology of flowering and fruiting are strongly related ($\chi^2 = 56.3$, $df = 4$, $P < 0.0000$; Fig. 2); consequently, floral preformation and fruiting season are also linked ($\chi^2 = 31.9$, $df = 2$, $P < 0.0000$; Fig. 1). For example, all species that fruit in the spring exhibit floral preformation, whereas smaller percentages of summer- and fall-fruiting species preform their flowers (Fig. 1).

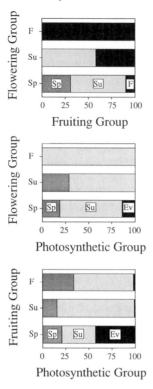

Fruiting Group

Photosynthetic Group

Photosynthetic Group

Figure 2   Associations between the seasons of flowering and fruiting, flowering and photosynthesis, and fruiting and photosynthesis. Data come from a survey of herbs in Wisconsin upland forests (Randall, 1952; see text). In each graph, the percentage of species in each phenological group on the x axis is plotted for each phenological group on y axis. Phenological groups as in Fig. 1.

There are also associations between floral preformation and the season of carbon gain ($\chi^2 = 5.49$, $df = 2$, $P < 0.07$), with preformation being more common in spring-active species compared to summer-active or evergreen species (Fig. 1). Finally, there are nonrandom associations between the timing of flowering and fruiting and the season of photosynthesis (flowering versus photosynthesis: $\chi^2 = 14.0$, $df = 4$, $P < 0.007$; fruiting versus photosynthesis: $\chi^2 = 28.1$, $df = 4$, $P < 0.0000$). Thus, species that flower and fruit in the spring may make their carbon gain at any time of year, whereas species that reproduce later in the year are rarely ever evergreen (Fig. 2).

These data show that there are strong recurring themes in the relationship between the phenology of development, reproduction, and carbon gain. Regardless of taxonomic affiliation, species that reproduce and are photosynthetically active early in the growing season are very likely to preform their flower buds (Fig. 1), whereas preformation is absent from species that flower in the fall. These patterns notwithstanding, there is also considerable diversity in how phenological traits cooccur. For instance, among summer- and fall-flowering species, some commit to flowering far in advance, while others do not (Fig. 1). In addition, depending on the species, flowering and fruiting can precede, coincide with, or follow the period of carbon gain (Fig. 2).

The implications of this diversity of phenology for variation in life history schedules and hence for demography are virtually unexplored. In view of the fact that a species' abundance and persistence at a site are determined, in a proximate sense, by its demographic parameters (Caswell, 1989), and that the latter are themselves an outcome of developmental events, it means that the phenology of development may have far-reaching consequences for species' distributions. Thus, the link between phenology and demography and the fact that phenological characteristics of forest herbs are phylogenetically conserved may explain why there is also a strong phylogenetic component to the geographic ranges of forest herbs (Ricklefs, 1989; Ricklefs and Latham, 1992).

To better understand the connections between phenology, life history, demography, and ultimately plant distribution, more precise information will be needed on the timing of development and of resource allocation and expenditure. Missing from Randall's study, for example, is more detailed information on the exact timing of flower preformation. Are flowers preformed in the fall, at the end of the prior growing season when plants have complete information about the season's resource gains; or are meristem fates decided much earlier, before plants have fully invested in the prior season's flowering and fruiting and before resource gains are complete?

III. Demographic Implications of Developmental Variation

In the remainder of this chapter, we explore the demographic consequences of variation in the timing of developmental events in mayapple, a common spring-flowering herb in the understory of mesic forests in eastern North America. In mayapple, flowers and vegetative structures are entirely preformed in the year prior to their emergence, and the developmental decisions concerning next year's shoots are made before the current year's shoots have completed their reproduction and carbon gain. Watson and co-workers have been engaged in long-term studies of mayapple with the aim of examining the interaction between the plant's developmental program and patterns of resource uptake and use, and of understanding the demographic consequences of this interplay (Benner and Watson, 1989;

Watson, 1990; de Kroon *et al.*, 1991; Landa *et al.*, 1992; Geber *et al.*, 1997; Lu, 1996; C. S. Jones and M. A. Watson, unpublished manuscript, 1996). We briefly describe the phenology of resource distribution and of development in mayapple and the interaction that takes place between the current year's reproduction and the developmental decisions concerning next year's shoots. It is this interaction that determines the manifestation of short-term demographic costs of reproduction in mayapple. We then use matrix simulations to examine the long-term demographic consequences of variation in the developmental phenology, as mediated through the costs of reproduction.

A. The Phenology of Shoots, Resource Distribution, and Development in Mayapple

Mayapples maintain long-lived rhizome systems composed of a series of annual rhizome segments, or sympodial units, that remain interconnected for several years (Fig. 3). Each year, an annual aerial shoot is borne on the terminal (youngest) rhizome segment and is either vegetative or sexual (Foerste, 1884; Holm, 1899; Martin, 1958; DeMaggio and Wilson, 1986).

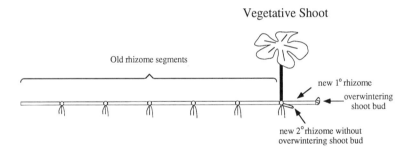

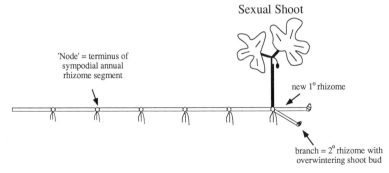

Figure 3 Schematic of mayapple aerial shoots and rhizomes. In a given year, the aerial shoot can be either vegetative (top) or sexual (bottom).

A system's shoot type, and thus its demographic status, can change from year to year.

In southern Indiana, the site of most of our studies (Landa *et al.*, 1992), shoots usually emerge in late March to early April and senesce between June and September (Lu, 1996). Sexual systems (i.e., systems whose current aerial shoot is sexual) flower in May and mature fruit in June to July (Fig. 4). Studies on the movement of ^{14}C-labeled photosynthate and experiments that sever the rhizome show that there is a seasonal pattern to the transport of carbon resources. Expansion of the current shoot draws on carbon stores in older rhizome segments (Landa *et al.*, 1992; M. A. Geber and M. A. Watson, unpublished data, 1994; Fig. 4). Carbon fixed early in the current season is primarily retained in the aerial shoot; carbon fixed near the time of flowering moves back into older rhizome segments, whereas carbon fixed later in the season moves into new rhizome segments (Landa *et al.*, 1992; Fig 4). Ongoing studies of carbon movement are examining the carryover of carbon fixed in one year to structures developed in subsequent years. The possibility of such carryover effects is suggested by the fact that a plant's history of shoot types affects the demographic status of a plant several years hence (Lu, 1996; Geber *et al.*, 1997).

While the current shoot is active, two developmental events affecting next year's demography are taking place below ground: the decision by the rhizome of whether or not to branch, and the determination of next year's shoot type. In early May, following expansion of the current shoot, one or more new sympodial rhizome segments begin elongating from lateral rhizome buds. Branching occurs when two or more of these segments elongate to produce preformed shoot buds containing either a vegetative or a sexual shoot (Fig. 3). More typically, the smaller new rhizome segments cease growing at an early stage and do not produce shoot buds. In these systems, only the largest new segment produces a preformed shoot bud, and the rhizome system does not branch. Our studies indicate that the commitment to branching takes place quite early in the season, most likely by mid to late May, before fruits have begun to develop on sexual systems and well before the current shoots senesce (Fig. 4; Geber *et al.*, 1997).

The differentiation of shoot buds on the new rhizome segments occurs later, in mid to late June; at this time sectioned terminal buds reveal recognizable vegetative or sexual shoots (Lu, 1996; C. S. Jones and M. A. Watson, unpublished manuscript, 1996). Thus, the determination of next year's shoot type takes place while reproductive systems are investing in fruit, but generally before the current shoots senesce (Fig. 4; Geber *et al.*, 1997).

Studies show that developmental events concerning branching and shoot type are affected by a plant's resource status; resource status, in turn, is a

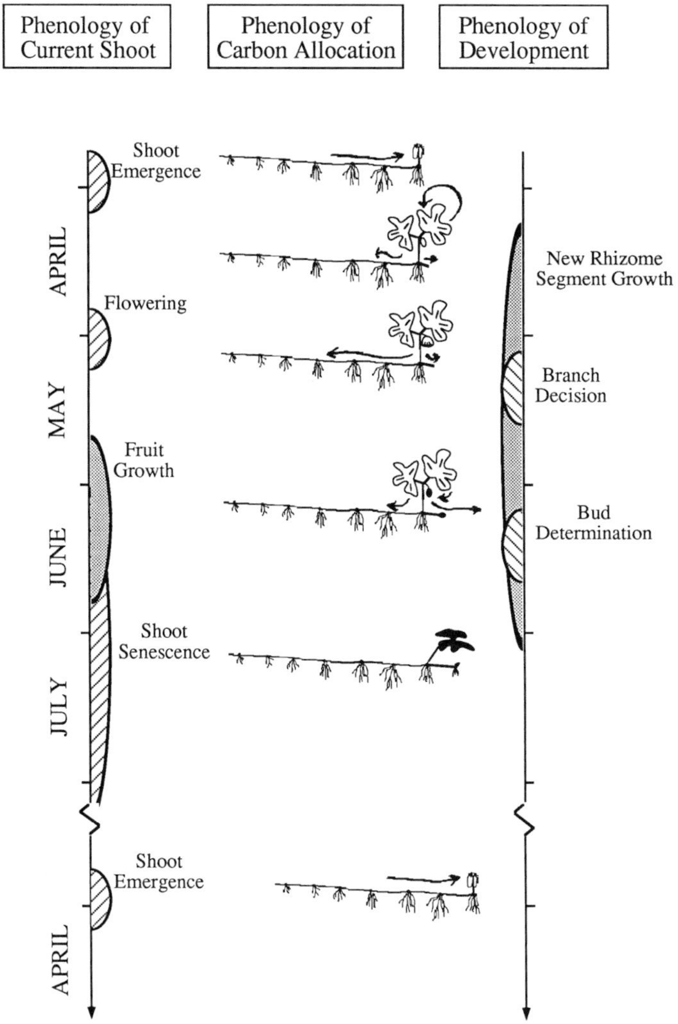

Figure 4 Seasonal phenology of mayapple. Events pertaining to the current shoot are depicted on the left-hand axis, whereas developmental events below ground are depicted on the right-hand axis. The phenology of carbon translocation is shown in the middle diagrams of mayapple systems.

function of a plant's ability to acquire resources—as determined by its leaf area—and its resource expenditures (Sohn and Policansky, 1977; Geber *et al.*, 1997). In particular, the resource costs of reproduction exact a demographic cost on new shoot production, but the nature of these costs is related to the timing of development.

B. Preformation and the Demographic Costs of
Reproduction in Mayapple

A key factor in the evolution of life histories is the trade-off between current reproduction and subsequent growth, reproduction, and survival (Cole, 1954; Gadgil and Bossert, 1970; Schaffer and Gadgil, 1975). This trade-off is often referred to as the demographic cost of reproduction (Bell, 1985; Reznick, 1985). In plants, short-term demographic costs are typically expressed through the initiation of fewer or smaller new leaves, shoots, and flowers following an episode of reproduction; these demographic costs are mediated through changes in developmental events affecting the numbers, sizes, and types of new organs (Law, 1979; Watson, 1984; Montalvo and Ackerman, 1987; Paige and Whitham, 1987; Clark and Clark, 1988; Snow and Whigham, 1989; Zimmerman and Aide, 1989; Ackerman and Montalvo, 1990; Karlsson *et al.*, 1990; Primack and Hall, 1990; Newell, 1991; Fox and Stevens, 1991; Eggert, 1992; Galen, 1993; Méndez and Obeso, 1993; Obeso, 1993; Syrjänen and Lehtilä, 1993; Primack *et al.*, 1994). In order for reproduction to affect developmentally mediated demographic parameters, the developmental events concerning new organ production must take place after, or at the very least at the same time as, investments in reproduction. Reproductive investments cannot affect developmental events that take place before reproductive costs are incurred, other than to cause subsequent abortion of organs.

In mayapple, the timing of developmental events affects the expression of short-term demographic costs (Reznick, 1985, 1992). For example, the decision to branch which takes place early in the season—before fruits are matured (Fig. 4)—is not affected by fruiting. Indeed, experimental removals of growing fruit and/or hand pollinations that enhance fruit set have had no effect on branching at our study site in Indiana or in a population near Ithaca, New York (Table III; Geber *et al.*, 1997). By contrast, the determination of new shoots, which takes place later in the season (Fig. 4), is affected by fruiting. Thus, sexual systems that bear fruit at our study site are less likely to produce new sexual shoot buds, and are more likely to produce vegetative shoot buds, compared to nonfruiting sexual systems (Geber *et al.*, 1997; Lu, 1996). Typically, fruiting sexuals resemble vegetative systems with respect to new shoot bud fates (Geber *et al.*, 1997). Shoot bud fates in sexual systems remain responsive to a system's reproductive status, and hence to the resource demands of reproduction, at least until late May (i.e., 2–3 weeks postpollination). When growing fruits are experimentally removed from fruiting systems at this time, shoot bud fates on new rhizome segments are similar to those of nonfruiting sexuals (Table IV; Geber *et al.*, 1997). However, when fruit is removed 6 weeks after pollination, a system's chance of producing new sexual shoot buds is identical to that of

Table III Frequency of Branching in Sexual Mayapples as a Function of Current Reproductive Status in Populations from Southern Indiana and Central New York[a]

Population	Flower loss[b]	Reproductive status		
		Fruit loss[c,d] (3 weeks)	Fruit loss[c,e] (6 weeks)	Fruiting[c]
Southern Indiana	28	21	—	26
Central New York	15	27	16	25

[a] In neither population does the reproductive status of sexual systems affect the likelihood of branching. Indiana data are from Geber *et al.* (1997).

[b] In the Indiana population, flower loss occurred through natural abortion of flower buds as shoots were expanding very early in the season, possibly in response to cold weather (Geber *et al.*, 1997). In the New York population, flower buds were experimentally aborted at the time of shoot expansion.

[c] In both populations, all sexual systems that retained flowers were hand pollinated to assure fruit set.

[d] In both populations, fruit loss was due to experimental removal of fruit approximately 2–3 weeks after pollination of flowers.

[e] In the New York population, fruit was removed experimentally from a second set of sexual systems 6 weeks after pollination, when fruits had attained approximately two-thirds the size of mature fruits.

fruiting sexuals (Table IV; see data on New York population). These results are consistent with the notion that developmental decisions—in this case branching—are only affected by reproductive investments that precede or are contemporaneous with developmental events.

C. Variation in Developmental Phenology and Its Demographic Implications

Sohn and Policansky (1977) reported a different result for two mayapple populations, one in northern Indiana and another in Tennessee. At their

Table IV Frequency of New Sexual Shoot Buds in Sexual Mayapples as a Function of Current Reproductive Status in Populations from Southern Indiana and Central New York[a]

Population	Flower loss	Reproductive status		
		Fruit loss (3 weeks)	Fruit loss (6 weeks)	Fruiting
Southern Indiana	69[a]	63[a]	—	31[b]
Central New York	48[a]	47[a]	15[b]	13[b]

[a] Within each population, different letter superscripts indicate significant differences in the frequency of sexual shoot formation among plants of different reproductive status. Reproductive status designations are as in Table III.

sites, fruiting had a negative effect on branching. Thus, fruiting sexuals were much less likely to branch (40–49%) compared to nonfruiting sexuals (71–79%). We suggest that population differences in the effect of fruiting on branching could arise if there is geographic variation in the relative timing of new rhizome growth and fruit set. In particular, if, in some populations, new rhizome growth and the decision to branch are delayed relative to fruit development (Fig. 5, right-hand side), then branching might decrease in response to reproduction, as at the sites studied by Sohn and Policansky. Variation in developmental phenology can also have important consequences for the long-term demography of plants. In the case of mayapple, where branching is the means by which plants spread clonally, genets experiencing similar levels of fruiting could nonetheless

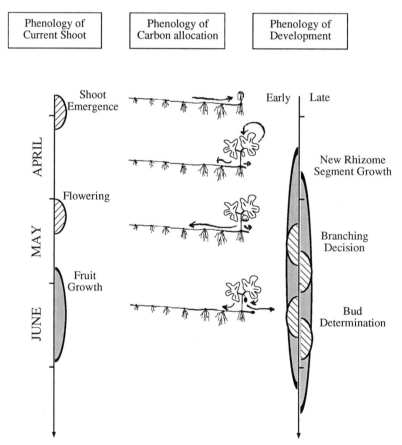

Figure 5   Seasonal phenology of mayapple, illustrating an early and a late developmental phenology of new rhizome segment growth and branching.

differ in the degree of clonality (i.e., the degree to which shoots are connected below ground) and in the rate of shoot population growth, simply because of variation in the phenology of new rhizome growth and branching. Additionally, genets may also differ in the proportion of their shoots that are sexual because of the influence of fruiting on shoot bud fates.

To illustrate the possible effects of variation in developmental phenology on long-term demography, we use matrix simulations to make projections about the growth of shoot numbers and the frequency of sexual shoots in genets (Sohn and Policansky, 1977; Caswell, 1989).

IV. Matrix Model

In any given year, a mayapple genet can be viewed as a collection of shoots located at the terminal ends of all of the genet's rhizomes. The shoots are of two types, vegetative and sexual; moreover, sexual shoots may or may not fruit (Sohn and Policansky, 1977). To project the number of new sexual and vegetative shoots in a genet in year $t + 1$, it is necessary to know the expected numbers of new sexual shoots and new vegetative shoots produced by an extant sexual or vegetative shoot in year t:

$$\begin{bmatrix} n_s \\ n_v \end{bmatrix}_{t+1} = \begin{bmatrix} a_{11} & a_{12} \\ a_{21} & a_{22} \end{bmatrix} \cdot \begin{bmatrix} n_s \\ n_v \end{bmatrix}_t \qquad (1)$$

where n_s and n_v are the numbers of sexual and vegetative shoots in a genet and a_{ij} is the expected number of new shoots of type i produced per shoot of type j. Using vector notation, this can be rewritten as

$$\mathbf{n}_{t+1} = \mathbf{A} \cdot \mathbf{n}_t \qquad (2)$$

where $\mathbf{n}$ is a 2×1 vector of shoots and $\mathbf{A}$ is 2×2 transition matrix whose elements are the a_{ij} values. Assuming that the a_{ij} values remain constant from year to year, the dominant eigenvalue, λ, of $\mathbf{A}$ measures the growth rate of shoot numbers in a genet, once the shoot population has achieved a stable stage distribution, i.e., a stable ratio of sexual to vegetative shoots. The stable stage distribution is given by the dominant right eigenvector, $\mathbf{w}$, of $\mathbf{A}$ (Caswell, 1989) and can be used to calculate the stable frequency of sexual shoots in a genet. If w_s and w_v are the elements of $\mathbf{w}$, then the stable frequency of sexual shoots in a genet is simply $w_s/(w_s + w_v)$.

The a_{ij} values can be estimated from demographic data on the likelihood that a sexual/vegetative shoot produces 0, 1, 2, or more new sexual/vegetative shoots per year (Sohn and Policansky 1977). Calculations of the a_{ij} values require data on the following demographic parameters for vegetative, fruiting, and nonfruiting sexual shoots: (1) shoot mortality, Q (i.e., the probability that a shoot dies and fails to produce any new shoots); (2) for shoots that do not die, the probability of branching, B; (3) for shoots that do not branch, the probabilities P and $(1 - P)$ of forming a sexual or a vegetative shoot bud; and (4) for shoots that branch, the probabilities R, S, and T of forming 2 sexual buds; 1 sexual and 1 vegetative bud; or 2 vegetative buds $(R + S + T = 1)$. (In our models we limit branching to the production of two new shoots.) It is also necessary to know, for sexual shoots, the probability of fruiting (F). In mayapple F is likely to be a function of pollinator abundance, as fruit set appears to be pollinator-limited, at least within years (Sohn and Policansky, 1977; Laverty and Plowright, 1988; Laverty, 1992; Whisler and Snow, 1992). Using the subscripts v, f, and nf to reference probabilities for vegetative, fruiting, and nonfruiting shoots, the matrix elements, a_{ij}, are as follows:

$$a_{11} = (1 - F)(1 - Q_{nf})[B_{nf}(2R_{nf} + S_{nf}) + (1 - B_{nf})P_{nf}] \\ + F(1 - Q_f)[B_f(2R_f + S_f) + (1 - B_f)P_f] \tag{3a}$$

$$a_{21} = (1 - F)(1 - Q_{nf})[B_{nf}(S_{nf} + 2T_{nf}) + (1 - B_{nf})(1 - P_{nf})] \\ + F(1 - Q_f)[B_f(S_f + 2T_f) + (1 - B_f)(1 - P_f)] \tag{3b}$$

$$a_{12} = (1 - Q_v)[B_v(2R_v + S_v) + (1 - B_v)P_v] \tag{3c}$$

$$a_{22} = (1 - Q_v)[B_v(S_v + 2T_v) + (1 - B_v)(1 - P_v)]. \tag{3d}$$

Sohn and Policansky (1977) used this model to examine the long-term demographic costs of reproduction in mayapple, by comparing the demography of genets in which sexual shoots always fruit (i.e., $F = 1$ and reproductive costs are always present) to the demography of genets in which sexual shoots never fruit (i.e., $F = 0$ and reproductive costs are always absent).

Our interest is in how the phenology of branch development affects genet demography. As noted earlier, branching phenology influences the demographic parameters of fruiting and nonfruiting sexuals in Eqs. (3a) and (3b), through its effect on the expression of demographic costs to reproduction. Thus, if fruiting has a negative effect on branching, because the decision to branch coincides with investments in the fruit, then $B_f <$ B_{nf}. Likewise, if fruiting reduces the likelihood of forming new sexual shoots, then $P_f < P_{nf}$ and/or $R_f < R_{nf}$.

We lack the data to estimate all of the demographic parameters in Eqs. (3a)–(3d) and do not attempt therefore to make projections about real genets. Instead, we simulate the effects branching phenology on genet demography by constructing two matrices, one for a genet with an early-season branching decision and the other for a genet with a late-season branching decision. The two genets are referred to hereafter as early and late genets.

A. Model Assumptions

For early genets, we assume that branching is not affected by reproduction ($B_f = B_{nf}$; southern Indiana and New York populations) but that shoot bud fates are responsive to fruiting ($P_f < P_{nf}$, $R_f < R_{nf}$, $S_f < S_{nf}$). Our data on unbranched systems indicate that new shoot fates do differ between fruiting and nonfruiting sexuals (i.e., $P_f < P_{nf}$; Table III), but, at present, we do not have adequate data to evaluate the impact of fruiting on new shoot fates in branched systems.

For late genets, we assume that branching is affected by reproduction ($B_f < B_{nf}$; populations of Sohn and Policansky). We also assume, as did Sohn and Policansky (1977; p. 1371), that fruiting and nonfruiting sexuals have identical new shoot bud fates (i.e., $P_f = P_{nf}$, $R_f = R_{nf}$, $S_f = S_{nf}$). This last assumption may not be fully correct, because fruiting sexuals in the populations studied by Sohn and Policansky produced smaller new rhizome segments than nonfruiting sexuals, and smaller segments are more likely to bear vegetative shoots (Sohn and Policansky, 1977).

In summary, in our models, an early genet adjusts to the costs of fruiting by altering the type of new shoots produced, whereas a late genet adjusts to these costs by reducing shoot number. These simplified assumptions probably do not capture the full complexity of developmental responses in mayapple, and our purpose in using them is simply to illustrate the potential impact of developmental variation on long-term demography. Furthermore, these assumptions minimize the demographic differences between early and late genets because, although distinct in their responses, the two genets have counterbalancing ways of absorbing the costs of reproduction.

We also assume, in our models, that when fruiting and nonfruiting sexuals differ from one another in branching frequency or in new shoot fates, then fruiting sexuals and vegetative shoots have the same demographic parameters (i.e., $B_f = B_v < B_{nf}$ or $P_f = P_v < P_{nf}$ and $R_f = R_v < R_{nf}$; Sohn and Policansky, 1977; Geber *et al.*, 1997; see Section III,B). For simplicity, we assume that new shoot fates are the same for branched and unbranched systems (i.e., $P_{nf} = R_{nf} + 0.5 S_{nf}$ and $P_v = R_v + 0.5 S_v$). Finally, we ignore possible differences among shoot types in mortality (i.e., $Q_f = Q_{nf} = Q_v$); mortality can then be dropped from Eq. (3a)–(3d).

The matrix elements a_{11} and a_{21} for the early genet reduce to the following:

$$a_{11} = (1 - F)(1 + B_{nf})P_{nf} + F(1 + B_{nf})P_v \tag{4a}$$

$$a_{21} = (1 - F)(1 + B_{nf})(1 - P_{nf}) + F(1 + B_{nf})(1 - P_v), \tag{4b}$$

and for the late genet we have

$$a_{11} = (1 - F)(1 + B_{nf})P_{nf} + F(1 + B_v)P_{nf} \tag{5a}$$

$$a_{21} = (1 - F)(1 + B_{nf})(1 - P_{nf}) + F(1 + B_v)(1 - P_{nf}) \tag{5b}$$

The first terms on the right-hand side of Eqs. (4) and (5) represent the contribution of nonfruiting sexuals to new shoot production and does not differ between early and late genets. The second terms represent the contribution of fruiting sexuals to next year's shoots and does differ between early and late genets. The contributions of vegetative shoots to new shoot production (a_{12} and a_{22}) are identical for both genets, and are given by Eqs. (3c) and (3d), except that the mortality term $(1 - Q_v)$ has been eliminated.

In our simulations, we examine the demographic consequences of varying (1) fruit set (F); (2) the frequency of branching, B_{nf}, and (3) the difference in branching between nonfruiting sexuals and vegetative shoots ($B_{nf} = 3B_v$, $2B_v$, or B_v); (4) the frequency of new sexual shoot production (P_{nf}); and (5) the difference in new sexual shoot production between nonfruiting sexuals and vegetative shoots ($P_{nf} = 3P_v$, $2P_v$, or P_v).

B. Model Results

Null Case: $B_{nf} = B_v$ and $P_{nf} = P_v$

When branching and new sexual shoot production are the same for nonfruiting sexuals and vegetative shoots, then all shoots, including fruiting sexuals, have the same demographic parameters, and developmental phenology obviously has no effect on a genet demography. A genet's rate of shoot population growth, λ, is solely a function of the frequency of branching, B_{nf}, and does not depend on new sexual shoot production, P_{nf}, or on fruit set, F (Fig. 6). Because there is no shoot mortality in our model,

$$B_{nf} = B_v \quad \text{and} \quad P_{nf} = P_v$$

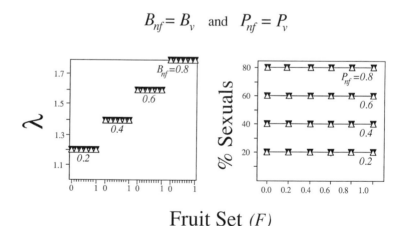

Fruit Set *(F)*

Figure 6 Influence of fruit set (F) and probabilities of branching (B_{nf}) and of sexual shoot formation (P_{nf}) on genet growth rate (λ) and on the frequency of sexual shoots (% sexuals) in a genet, under the assumption that branching and sexual shoot formation are the same in nonfruiting sexuals and vegetative shoots ($B_{nf} = B_v$ and $P_{nf} = P_v$). Under these assumptions early ($\triangle$) and late ($\blacktriangledown$) genets do not differ in growth rate or in shoot composition.

the size of the shoot population remains unchanged over time if there is no branching (i.e., $\lambda = 1$); with branching, λ is augmented by the frequency of branching (i.e., $\lambda = 1 + B_{nf}$), because each branching event contributes an additional new shoot (Fig. 6). The frequency of sexual shoots in a genet is simply equal to the probability of new sexual shoot formation, P_{nf}, and is unaffected by branching or by fruit set (Fig. 6).

Nonfruiting Sexuals and Vegetative Systems Differ in Branching: $B_{nf} < B_v$

Effects on λ. When the frequency of branching differs between nonfruiting sexuals and vegetative shoots, the growth rates of both early and late genets depend simultaneously on branching and on new sexual shoot formation, and growth rates increase as B_{nf} and P_{nf} increase. The dependence of λ on branching and sexual shoot formation is illustrated in Fig. 7 for the case where there is a 3-fold difference in branching between nonfruiting sexuals and vegetative shoots. A comparison of Figs. 7 and 6 shows that, even in the absence of fruiting, the growth rate of both genets is reduced relative to the null case (i.e., $\lambda < 1 + B_{nf}$) because vegetative shoots now branch less than sexual shoots.

In late genets, λ depends strongly on fruit set because fruiting reduces branching directly in these genets (Fig. 7, left-hand graphs). In contrast,

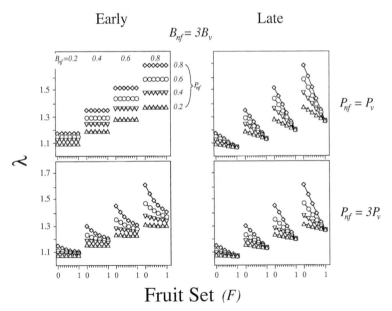

Figure 7 Influence of fruit set (F) and probabilities of branching (B_{nf}) and of sexual shoot formation (P_{nf}: $\triangle$, 0.2; $\triangledown$, 0.4; $\bigcirc$, 0.6; $\diamondsuit$, 0.8) on genet growth rate (λ) in early (left-hand graphs) and late genets (right-hand graphs), under the assumption that branching is three times higher in nonfruiting sexuals compared to vegetative shoots ($B_{nf} = 3B_v$).

in early genets, λ depends on fruit set only when nonfruiting sexuals also differ from vegetative shoots (and fruiting sexuals) in the probability of forming new sexual shoots (Fig. 7, right-hand graphs). The dependence of λ on fruit set is less pronounced in early genets (compare right- and left-hand sides of Fig. 7), because fruiting has only an indirect effect on shoot numbers in these genets: fruiting reduces the production of new sexual shoots, and hence the number of eventual branches, arising from fruiting sexuals.

Late genets always grow more slowly than early genets, except when there is no fruiting (i.e., $F = 0$; Fig. 7). One way to visualize the difference in λ between early and late genets is to look at the proportionate difference in growth rate between them [% difference $= 100(\lambda_{early} - \lambda_{late})/\lambda_{late}$]. As illustrated in Fig. 8, this difference in growth rate can be large, especially when the frequencies of branching, of new sexual shoot production, and of fruit set are high; and when nonfruiting sexuals and vegetative shoots differ greatly in the probability of branching (compare top and bottom graphs in Fig. 8). The difference in growth rate between early and late

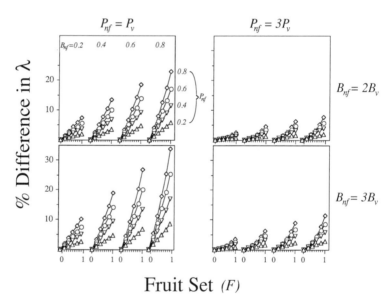

Fruit Set *(F)*

Figure 8 Proportionate difference in growth rate between early and late genets as a function of fruit set (F) and of the frequencies of branching (B_{nf}) and sexual shoot formation (P_{nf}: $\triangle$, 0.2; $\triangledown$, 0.4; $\bigcirc$, 0.6; $\diamond$, 0.8). A comparison of left- and right-hand graphs illustrates the dependence of genet differences in growth to new sexual shoot production in nonfruiting sexuals relative to vegetative shoots ($P_{nf} = P_v$ versus $P_{nf} = 3P_v$). A comparison of top and bottom graphs illustrates the dependence of genet differences in growth to branching in nonfruiting sexuals relative to vegetative shoots ($B_{nf} = B_v$ versus $B_{nf} = 3B_v$).

genets diminishes when nonfruiting sexuals and vegetative shoots differ in new sexual shoot production (compare right- and left-hand sides in Fig. 8), because, under these circumstances, early genets also experience a cost to reproduction, albeit more indirectly (Fig. 7).

Effects on Shoot Composition. The frequency of sexual shoots increases as a function of new sexual shoot production, P_{nf}, in both early and late genets (Fig. 9). Differences between vegetative shoots and nonfruiting sexuals in the formation of new sexual shoots have the effect of reducing the proportion of sexual shoots in a genet (compare top and bottom halves of Fig. 9) because vegetative shoots leave fewer sexual descendents. On the other hand, the shoot composition of a genet is not affected by the frequency of branching (in Fig. 9, B_{nf} is set at 0.6, but identical results are obtained with other values of B_{nf}), and it is only moderately affected by differences in branching between nonfruiting sexuals and vegetative shoots (compare right- and left-hand sides of Fig. 9).

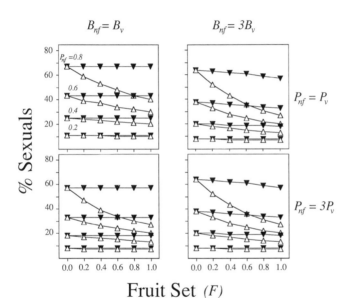

Fruit Set *(F)*

Figure 9 Dependence of shoot composition, as measured by the percentage of a genet's shoots that are sexual, on fruit set *(F)* and on the frequencies of branching (B_{nf}) and sexual shoot formation (P_{nf}) for early $(\triangle)$ and late $(\blacktriangledown)$ genets. A comparison of left- and right-hand graphs illustrates the dependence of shoot composition on branching in nonfruiting sexuals relative to vegetative shoots $(B_{nf} = B_v$ versus $B_{nf} = 3B_v)$. The actual frequency of branching in nonfruiting sexuals, which is set at $B_{nf} = 0.6$ for this figure, has no effect on shoot composition. A comparison of top and bottom graphs illustrates the dependence of shoot composition on new sexual shoot formation in nonfruiting sexuals relative to vegetative shoots $(P_{nf} = P_v$ versus $P_{nf} = 3P_v)$.

The shoot composition of early genets depends strongly on fruit set (Fig. 9, filled symbols). In contrast, fruit set has only a slight effect on shoot composition in late genets, and only when vegetative shoots and nonfruiting sexuals also differ in the probability of branching (compare right- and left-hand graphs in Fig. 9). The stronger dependence of shoot composition on fruit set in early genets is due to the fact that fruiting directly affects new sexual shoot formation in these genets, but does so only indirectly in late genets. As a result, the frequency of sexual shoots in early genets is always lower than that of late genets, except when there is no fruiting (Fig. 9).

In summary, our models examine the influence of developmental phenology on two aspects of a genet's demography, namely, its growth rate and shoot composition. These two aspects are important fitness components in that they represent the contrasting modes of propagation available to clonal plants (Cook, 1985). Growth rate is a measure of clonal spread, whereas shoot composition influences the potential for sexual reproduction. Our principal conclusions are as follows. (1) Genet growth rate is more strongly tied to the likelihood of branching than it is to sexual shoot formation, but shoot composition depends more strongly on sexual shoot production than on branching. (2) Insofar as the developmental phenology of a genet affects whether the demographic costs of reproduction are expressed via reductions in branching (late genets) versus alterations of shoot bud fates (early genets), then phenology can have profound consequences on genet demography. (3) The aspect of a genet's demography that is most affected by reproduction is a direct function of how it expresses demographic costs. For instance, growth rate is most sensitive to fruiting in late genets, whereas shoot composition is most sensitive to fruiting in early genets. (4) As a corollary to conclusion (3), each genet "outperforms" the other in the one aspect of its demography that is least sensitive to reproduction: early genets have higher growth rates and late genets have higher frequencies of sexual shoots.

The last two conclusions have important implications for the evolution of plant traits related to reproduction and of developmental phenology itself. Thus, conclusion (3) can be rephrased as follows: a given change in fruit set results in a larger change in the shoot composition of an early versus a late genet and in a larger change in the growth rate of a late versus an early genet. Because the effect of a change in fruit set on a fitness component (e.g., clonal growth rate or shoot composition) is a measure of the strength of natural selection on fruiting (Lande, 1980; van Groenendael *et al.*, 1988), we conclude that natural selection on traits related to fruit set occurs primarily via the contribution of clonal spread to total fitness in late genets, whereas in early genets selection is mediated through the effects of shoot composition (i.e., sexual reproduction) on total fitness.

In other words, developmental phenology can influence the nature and pathways of selection on plant traits affecting fruit set (e.g., attractiveness to pollinators, breeding system). Second, because early and late genets emphasize different modes of clonal propagation, the one asexual and the other sexual, developmental phenology can influence the genetic structure of clonal populations. Third, depending on the selective value of asexual versus sexual propagation, different developmental phenologies may evolve. Alternatively, if developmental phenology is fixed (genetically invariant) in a population, an evolutionary change in the phenology of resource acquisition and allocation (e.g., storage) may be the means by which populations respond to selection favoring a shift in asexual versus sexual propagation.

V. Conclusions

In plants, life history schedules of growth and reproduction reflect the combined expression of patterns of development and of resource allocation. This is because the developmental program of organ production directly influences the schedule of resource uptake and use, while the supply of resources affects developmental "decisions" concerning the fates of meristems that give rise to these organs. Even so, while development and resource allocation are inextricably linked, developmental and allocation decisions may be separated in time. Thus, in perennial species from seasonal environments, the commitment of meristems to organs often precedes, by several months to several years, the allocation of resources to build them. The temporal separation between organ formation, resource allocation, and organ elaboration raises the question of how plants make demographically important developmental commitments when they may not have acquired the resources to elaborate the organs and when they lack information about environmental conditions at the time of organ elaboration.

In this paper, we examine the prevalence, taxonomic affinity, and ecological distribution of flower preformation, and we explore the relationship between the phenology of development and of resource acquisition and use in forest herbs. We also examine the demographic and life history consequences of preformation.

A community-wide study by Randall (1952) on temperate forest herbs in Wisconsin indicates that floral preformation is extremely common, being found in 58% and 49% of angiosperm species from northern and southern Wisconsin forests respectively. Randall's study suggests that floral preformation is phylogenetically conservative, as are other phenological traits, such as the seasons of flowering, fruiting, and photosynthesis. In addition, there are strong statistical associations between developmental and phenological traits. For example, species that reproduce and are photosynthetically active in early spring are much more likely to preform their flower buds than are species that flower or photosynthesize in the summer or fall. These results suggest that the advantage of preformation lies in permitting rapid floral expansion in the spring when temperatures are generally low. These patterns notwithstanding, there is also considerable diversity in how the phenological traits co-occur. Depending on the species, reproduction can precede, coincide with, or follow the period of carbon gain. The implications of this diversity of phenology on variation in life history schedules and demography are virtually unexplored.

We explore the possible consequence of variation in developmental phenology on the demography of one forest herb—the mayapple—which preforms both vegetative and flowering structures almost a year before these organs expand to become functional. In this species, the developmental decisions affecting the fate of these structures take place while shoots of the current year are investing in flowers or fruit. Our studies indicate that the demographic expression of current reproductive costs is affected by the timing of developmental decisions relative to the timing of reproduction. In particular, current flowering and fruiting only affect developmental decisions that are contemporaneous with or that follow investments in reproduction. We use matrix simulations to explore the demographic consequences of variation in developmental phenology by contrasting the demography of a genet in which decisions concerning the numbers of shoots to emerge next year are made prior to the current year's investment in reproduction (early genet) to the demography of a genet in which these decisions are delayed and are contemporaneous with reproduction (late genet). We show that two important aspects of demography—the rate of shoot population growth (λ) and shoot composition (% of sexual shoots)—are strongly affected by developmental phenology. These two demographic characteristics represent the contrasting modes of propagation available to clonal plants: λ is a measure of the rate of clonal spread while shoot composition influences the potential for sexual reproduction. Thus, our simple models clearly demonstrate the impact of development on plant demography, on the evolution of plant traits affecting fruiting and clonal growth, and on the genetics of clonal plant populations. The models also suggest that there is still much fruitful exploration to be done on the joint evolution of phenological traits and resource dynamics in plants.

Acknowledgments

This work was supported by a New York State Hatch grant and a U.S. Department of Agriculture grant to M. A. Geber; by grants from the National Science Foundation and Indiana University to M. A. Watson; and by a Fulbright Fellowship and grants from the Netherlands Organization for Scientific Research, the Niels Stensen Foundation, and the Royal Netherlands Academy of Sciences to H. de Kroon. We thank Cynthia S. Jones for bringing Randall's work to our attention.

References

Ackerman, J. D., and Montalvo, A. M. (1990). Short- and long-term limitations to fruit production in a tropical orchid. *Ecology* **71**, 263–272.

Amir, S., and Cohen, D. (1990). Optimal reproductive efforts and the timing of reproduction of annual plants in randomly varying environments. *J. Theor. Biol.* **147**, 17–42.

Bell, G. (1985). The costs of reproduction and their consequences. *Am. Nat.* **116**, 45–76.

Benner, B. L., and Watson, M. A. (1989). Developmental ecology of mayapple: Seasonal patterns of resource distribution in sexual and vegetative rhizome systems. *Funct. Ecol.* **3**, 539–547.

Bierzychudek, P. (1982). Life histories and demography of shade-tolerant temperate forest herbs: A review. *New Phytol.* **90**, 757–776.

Bliss, L. C. (1971). Arctic and alpine plant life cycles. *Annu. Rev. Ecol. Syst.* **2**, 405–438.

Bloom, A. J., Chapin III, F. S., and Mooney, H. A. (1985). Resource limitation in plants: An economic strategy. *Annu. Rev. Ecol. Syst.* **16**, 363–392.

Brown, R. T., and Curtis, J. T. (1952). The upland conifer–hardwood forests of northern Wisconsin. *Ecol. Monogr.* **22**, 217–234.

Caraco, T., and Kelly, C. K. (1991). On the adaptive value of physiological integration in clonal plants. *Ecology* **72**, 81–93.

Caswell, H. (1989). "Matrix Population Models." Sinauer, Sunderland, Massachusetts.

Chiariello, N., and Roughgarden, J. (1984). Storage allocation in seasonal races of an annual plant: Optimal vs. actual allocation. *Ecology* **65**, 1290–1301.

Clark, D. B., and Clark, D. A. (1988). Leaf production and the cost of reproduction in the neotropical rain forest cycad, *Zamia skinneri*. *J. Ecol.* **76**, 1153–1163.

Clausen, J. J., and Kozlowski, T. T. (1970). Observations on growth of long shoots in *Larix laricina*. *Can. J. Bot.* **48**, 1045–1048.

Cohen, D. (1971). Maximizing final yield when growth is limited by time or by limiting resources. *J. Theor. Biol.* **33**, 299–307.

Cole, L. C. (1954). The population consequences of life history phenomena. *Q. Rev. Biol.* **29**, 103–137.

Cook, R. E. (1985). Growth and development in clonal plant populations. *In* "Population Biology and Evolution of Clonal Organisms" (J. B. C. Jackson, L. W. Buss, and R. E. Cook, eds.), pp. 259-296. Yale Univ. Press, New Haven, Connecticut.

Critchfield, W. B. (1960). Leaf dimorphism in *Populus trichocarpa*. *Am. J. Bot.* **47**, 699–711.

Curtis, J. T., and McIntosh, R. P. (1951). An upland forest continuum in the prairie–forest border region of Wisconsin. *Ecology* **32**, 476–496.

Dafni, A., Cohen, D., and Noy-Meir, I. (1981). Life-cycle variation in geophytes. *Ann. MO. Bot. Gard.* **68**, 652–660.

de Kroon, H., Whigham, D. F., and Watson, M. A. (1991). Developmental ecology of mayapple: Effects of rhizome severing, fertilization and timing of shoot senescence. *Funct. Ecol.* **5**, 360–368.

DeMaggio, A. E., and Wilson, C. L. (1986). Floral structure and organogenesis in *Podophyllum peltatum* (Berberidaceae). *Am. J. Bot.* **73**, 21–32.

Diggle, P. K. (1994). The expression of andromonoecy in *Solanum hirtum* (Solanaceae): Phenotypic plasticity and ontogenetic contingency. *Am. J. Bot.* **81**, 1354–1365.

Eggert, A. (1992). Dry matter economy and reproduction of a temperate forest spring geophyte, *Allium ursinum*. *Ecography* 15, 45–55.

Fisher, R. A. (1930). "The Genetical Theory of Natural Selection." Oxford Univ. Press, London.

Foerste, A. F. (1884). The may apple. *Bull. Torrey Bot. Club* 11, 62–64.

Foerste, A. F. (1891). On the formation of flower buds in sping-blossoming plants during the preceding summer. *Bull. Torrey Bot. Club* 18, 101–106.

Fox, J. F., and Stevens, G. C. (1991). Costs of reproduction in a willow: Experimental responses vs. natural variation. *Ecology* 72, 1013–1023.

Fulford, R. M. (1966). The morphogenesis of apple buds. III. The inception of flowers. *Ann. Bot.* 30, 207–219.

Gadgil, M., and Bossert, W. H. (1970). Life history consequences of natural selection. *Am. Nat.* 102, 52–64.

Galen, C. (1993). Cost of reproduction in *Polemonium viscosum:* Phenotypic and genetic approaches. *Evolution* 47, 1073–1079.

Geber, M. A. (1990). The cost of meristem limitation in *Polygonum arenastrum:* Negative genetic correlations between fecundity and growth. *Evolution* 44, 799–819.

Geber, M. A., de Kroon, H., and Watson, M. A. (1997). Organ preformation in mayapple as a mechanism for historical effects on demography. *J. Ecol.* 85, in press.

Gill, A. M. (1971). The formation growth and fate of buds of *Fraxinus americanus* L. in Central MA. *Harvard Forest Paper* 20, 1–16.

Givnish, T. J. (1987). Comparative studies of leaf form: Assessing the relative roles of selective pressures and phylogenetic constraints. *New Phytol.* 106. 131–160.

Goldschmidt, E. E., and Golomb, A. (1982). The carbohydrate balance of alternate-bearing citrus trees and the significance of reserves for flowering and fruiting. *J. Am. Soc. Hortic. Sci.* 107, 206–208.

Harper, J. L. (1989). The value of a leaf. *Oecologia* 80, 53–58.

Holm, T. (1899). *Podophyllum peltatum,* a morphological study. *Bot. Gaz.* 27, 419–433.

Inouye, D. W. (1986). Long-term preformation of leaves and inflorescences by a long-lived perennial monocarp, *Frasera speciosa* (Gentianaceae). *Am. J. Bot.* 73, 1535–1540.

Iwasa, Y. (1991). Pessimistic plant: Optimal growth schedule in stochastic environments. *Theoretical Population Biology* 40, 246–268.

Iwasa, Y., and Cohen, D. (1989). Optimal growth schedule of a perennial plant. *Am. Nat.* 133, 480–505.

Iwasa, Y., and Roughgarden, J. (1984). Shoot/root balance of plants: Optimal growth of a system with many vegetative organs. *Theoretical Population Biology* 25, 78–105.

Karlsson, P. S., Svensson, B. M., Carlsson, B. A., and Nordell, K. O. (1990). Resource investment in reproduction and its consequences in three *Pinguicula* species. Oikos 59, 393–398.

King, D., and Roughgarden, J. (1982a). Multiple switches between vegetative and reproductive growth in annual plants. *Theoretical Population Biology* 21, 194–204.

King, D., and Roughgarden, J. (1982b). Graded allocation between vegetative and reproductive growth for annual plants in growing seasons of random lengths. *Theoretical Population Biology* 22, 1–16.

Kochmer, J. P., and Handel, S. N. (1986). Constraints and competition in the evolution of flowering phenology. *Ecol. Monogr.* 56, 303–326.

Kozlowski, J. (1992). Optimal allocation of resources to growth and reproduction: Implications for age and size at maturity. *Trends Ecol. Evol.* 7, 15–19.

Kozlowski, J., and Uchmanski, J. (1987). Optimal individual growth and reproduction in perennial species with indeterminate growth. *Evol. Ecol.* 1, 214–230.

Kozlowski, J., and Wiegert, R. G. (1987). Optimal age and size at maturity in annuals and perennials with determinate growth. *Evol. Ecol.* 1, 231–244.

Kozlowski, J., and Ziolko, M. (1988). Gradual transition from vegetative to reproductive growth is optimal when the maximum rate of reproductive growth is limited. *Theoretical Population Biology* **34**, 118–129.

Lacey, E. P. (1986). Onset of reproduction in plants: Size- versus age-dependency. *Trends Ecol. Evol.* **1**, 72–75.

Landa, K., Benner, B., Watson, M. A., and Garner, J. (1992). Physiological integration for carbon in mayapple (*Podophyllum peltatum*), a clonal perennial herb. *Oikos* **63**, 348–356.

Lande, R. (1980). A quantitative genetic theory of life history evolution. *Ecology* **63**, 607–615.

Laverty, T. M. (1992). Plant interactions for pollinator visits: A test of the magnet species effect. *Oecologia* **89**, 502–508.

Laverty, T. M., and Plowright, R. C. (1988). Fruit and seed set in mayapple (*Podophyllum peltatum*): Influence of intraspecific factors and local enhancement near *Pedicularis canadensis. Can. J. Bot.* **66**, 173–178.

Law, R. (1979). The cost of reproduction in an annual meadowgrass. *Am. Nat.* **113**, 3–16.

Lechowicz, M. J. (1984). Why do temperate trees leaf out at different times? Adaptations and ecology of forest communities. *Am. Nat.* **124**, 821–842.

Lerdau, M. (1992). Future discounts and resource allocation in plants. *Funct. Ecol.* **6**, 371–375.

Lu, Y. (1996). Developmental ecology of mayapple, *Podophyllum peltatum* (Berberidaceae). Ph.D. Thesis, Indiana University, Bloomington.

Marino, F., and Greene, D. W. (1981). Involvement of gibberellins in the biennial bearing of 'Early MacIntosh' apples. *J. Am. Soc. Hortic. Sci.* **106**, 593–596.

Marks, P. L. (1975). On the relation between extension growth and successional status of deciduous trees of the Northeastern United States. *Bull. Torrey Bot. Club* **102**, 172–177.

Martin, F. W. (1958). Variation and morphology of *Podophyllum peltatum*. Ph.D. Thesis, Washington University, St. Louis, Missouri.

Méndez, M., and Obeso, J. R. (1993). Size-dependent reproductive and vegetative allocation in *Arum italicum* (Araceae). *Can. J. Bot.* **71**, 309–314.

Merrill, E. K. (1990). Structure and development of terminal bud scales in green ash. *Can. J. Bot.* **68**, 12–20.

Montalvo, A. M., and Ackerman, J. D. (1987). Limitations to fruit production in *Ionopsis utricularioides* (Orchidaceae). *Biotropica* **19**, 24–31.

Moore, E. (1909). The study of winter buds with reference to their growth and leaf content. *Bull. Torrey Bot. Club* **37**, 117–145.

Newell, E. A. (1991). Direct and delayed costs of reproduction in *Aesculus californica. J. Ecol.* **79**, 365–378.

Nobel, P. S. (1987). Water relations and plant size aspects of flowering for *Agave deserti. Bot. Gaz.* **148**, 79–84.

Obeso, J. R. (1993). Cost of reproduction in the perennial herb *Asphodelus albus* (Liliaceae). *Ecography* **16**, 365–371.

Owens, J. N., and Molder, M. (1973a). A study of DNA and mitotic activity in the vegetative apex of Douglas-fir during the annual growth cycle. *Can. J. Bot.* **51**, 1395–1409.

Owens, J. N., and Molder, M. (1973b). Bud development in western hemlock. I. Annual growth cycle of vegetative buds. *Can. J. Bot.* **51**, 2223–2231.

Owens, J. N., and Molder, M. (1979). Bud development in *Larix occidentalis*. I. Growth and development of vegetative long shoot and vegetative short shoot buds. *Can. J. Bot.* **57**, 687–700.

Paige, K. N., and Whitham, T. G. (1987). Flexible life history traits: Shifts by scarlet gilia in response to pollinator abundance. *Ecology* **68**, 1691–1695.

Parke, B. (1959). Growth periodicity and the shoot tip of *Abies concolor. Am. J. Bot.* **46**, 110–118.

Perrin, N., and Sibly, R. M. (1993). Dynamic models of energy allocation and investment. *Annu. Rev. Ecol. Syst.* **24**, 379–410.

Primack, R. B., and Hall, P. (1990). Costs of reproduciton in the pink lady's slipper orchid: A four-year experimental study. *Am. Nat.* **136**, 638–656.

Primack, R. B., Miao, S. L., and Becker, K. B. (1994). Costs of reproduction in the pink lady's slipper orchid (*Cypripedium acaule*): Defoliation, increased fruit production, and fire. *Am. J. Bot.* **81**, 1083–1090.

Pugliese, A. (1987). Optimal resource allocation and optimal size in perennial herbs. *J. Theor. Biol.* **126**, 33–49.

Pugliese, A. (1988). Optimal resource allocation in perennial plants: A continuous-time model. *Theoretical Population Biology* **34**, 215–247.

Pugliese, A., and Kozlowski, J. (1990). Optimal patterns of growth and reproduction in perennial plants with persisting and not persisting vegetative parts. *Evol. Ecol.* **4**, 75–89.

Randall, W. J. (1952). Interrelations of autecological characteristics of forest herbs. Ph.D. Thesis. University of Wisconsin, Madison.

Raunkiaer, C. (1934). "The Life Forms of Plants and Statistical Plant Geography." Oxford Univ. Press (Clarendon), Oxford.

Reznick, D. (1985). Costs of reproduction: An evaluation of empirical evidence. *Oikos* **44**, 257–267.

Reznick, D. (1992). Measuring the costs of reproduction. *Trends Ecol. Evol.* **7**, 42–45.

Ricklefs, R. E. (1989). Speciation and diversity: The integration of local and regional processes. *In* "Speciation and Its Consequences" (D. Otte and J. A. Endler, eds.), pp. 599–622. Sinauer, Sunderland, Massachusetts.

Ricklefs, R. E., and Latham, R. E. (1992). Intercontinental correlation of geographic ranges suggests stasis in ecological traits of relict genera of temperate perennial herbs. *Am. Nat.* **139**, 1305–1321.

Sacher, J. A. (1954). Structure and sexual activity of the shoot apices of *Pinus lambertiana* and *Pinus ponderosa. Am. J. Bot.* **41**, 749–759.

Sattler, R. (1973). "Organogenesis of Flowers." Univ. Toronto Press, Toronto.

Schaffer, W. M., and Gadgil, M. D. (1975). Selection for optimal life histories in plants. *In* "Ecology and Evolution of Communities" (M. L. Cody and J. M. Diamond, eds.), pp. 142–157. Belknap Press, Cambridge, Massachusetts.

Schaffer, W. M., Inouye, R. S., and Whittam, T. S. (1982). Energy allocation by an annual plant when the effects of seasonality on growth and reproduction are decoupled. *Am. Nat.* **120**, 787–815.

Snow, A. A., and Whigham, D. F. (1989). Costs of flower and fruit production in *Tipularia discolor* (Orchidaceae). *Ecology* **70**, 1286–1293.

Sohn, J. J., and Policansky, D. (1977). The costs of reproduction in *Podophyllum peltatum* (Berberidaceae). *Ecology* **58**, 1366–1374.

Sørensen, Th. (1941). Temperature relations and phenology of the Northeast Greenland flowering plants. *Medd. Groenl.* **125**, 1–305.

Steeves, M. W., and Steeves, T. A. (1990). Inflorescence development in *Amelanchier alnifolia. Can. J. Bot.* **68**, 1680–1688.

Syrjänen, K., and Lehtilä, K. (1993). The cost of reproduction in *Primula veris*: Differences between two adjacent populations. *Oikos* **67**, 465–472.

Takeda, F., Ryugo, K., and Crane, J. C. (1980). Translocation and distribution of ^{14}C-photosynthates in bearing and nonbearing pistachio branches. *J. Am. Soc. Hortic. Sci.* **105**, 642–644.

van Groenendael, J., de Kroon, H., and Caswell, H. (1988). Projection matrices in population biology. *Trends Ecol. Evol.* **3**, 264–269.

Vincent, T. L., and Pulliam, H. R. (1980). Evolution of life-history strategies for an asexual annual plant model. *Theoretical Population Biology* **17**, 215–231.

Watson, M. A. (1984). Developmental constraints: Effect on population growth and patterns of resource allocation in a clonal plant. *Am. Nat.* **123**, 411–426.

Watson, M. A. (1990). Phenological effects on clone development and demography. *In* "Clonal Growth in Plants: Regulation and Function" (J. van Groenendael and H. deKroon, eds.), pp. 43–55. SPB Academic Publ., The Hague.

Watson, M. A., and Casper, B. B. (1984). Morphogenetic constraints on patterns of carbon distribution in plants. *Annu. Rev. Ecol. Syst.* **15**, 233–258.

Westwood, M. N. (1993). "Temperate-Zone Pomology: Physiology and Culture." Timber Press, Portland, Oregon.

Whisler, S. L., and Snow, A. A. (1992). Potential for the loss of self-compatibility in pollen-limited populations of mayapple (*Podophyllum peltatum*). *Am. J. Bot.* **79**, 1273–1278.

Yoshie, F., and Yoshida, S. (1987). Seasonal changes in photosynthetic characteristics of *Anemone raddeana*, a spring-active geophyte, in the temperate region of Japan. *Oecologia* **72**, 202–206.

Yoshie, F., and Yoshida, S. (1989). Wintering forms of perennial herbs in the cool temperate regions of Japan. *Can. J. Bot.* **67**, 3563–3569.

Zimmerman, J. K., and Aide, T. M. (1989). Patterns of fruit production in a neotropical orchid: Pollinator vs. resource limitation. *Am. J. Bot.* **76**, 67–73.

6

Optimality Approaches to Resource Allocation in Woody Tissues

R. M. Sibly and J. F. V. Vincent

I. Introduction

In recent years the principles of optimal resource allocation have given insight into the evolution of a wide range of plant characteristics, including the evolution of seed size (e.g., Sakai, 1995a; McGinley and Charnov, 1988); allocation between vegetative and reproductive tissue (Fox, 1992; Hara *et al.*, 1988; Pugliese, 1987) and between those and storage (Pugliese, 1988; also see review in Perrin and Sibly, 1993); allocation between biomass production and defense (Fagerstrom *et al.*, 1987) and between ramets, rhizomes, and seeds (Sakai, 1995b); the evolution of selfing (Iwasa, 1990; Sakai, 1995c); of floral longevity (Schoen and Ashman, 1995); of capsule-to-flower ratios (Bartareau, 1995), of parent–offspring conflict over the allocation of resources to seeds (Ravishankar *et al.*, 1995; Mazer, 1987), and of perennial growth (Iwasa and Cohen, 1989). Here we analyze as far as possible some cases of resource allocation to woody tissue, and then review the obstacles to obtaining a complete life-history analysis of this process.

A full understanding of resource allocation requires answers to five questions. To what are resources allocated at any particular stage or age? What physiologically controls resource allocation? How does resource allocation change during development? What is the survival value (fitness) of the observed resource allocation? And what is the evolutionary history (phylogeny) of the observed pattern of resource allocation? All of these are important. In this chapter, however, we deal mainly with the fitness implications of different patterns of resource allocation. This is a necessary part of life-history analysis. If all the fitness implications of each allocation strategy are known, then it is a simple matter to identify the most successful strategy in each environment.

We begin by describing what is needed in a complete life-history analysis and then go on to identify two especially simple and experimentally tractable cases for the study of allocation trade-offs. The first is an analysis of early seedling growth and a consideration of the many different ways in which different species may defend themselves, for which much data are available (Grime and Hunt, 1975). Here we show how energy allocation principles can provide a good theoretical basis for further analysis of seedling growth, and how this might be used to calculate the energy costs of different types of seedling defense.

In the second example we consider whether, and how, energy is best allocated in a more complex structure. Although we wished to consider a structure with some complexity, we also wanted one whose analysis would be tractable. We have therefore chosen to analyze the shapes of the horizontal branches of trees.

Studies of resource allocation assume a special importance in life-history analysis, because they can provide information in an area where data can be hard to come by, namely, identifying the shapes of the trade-off curves that limit the genetic options open to the organism in a particular study environment. This is a tantalizing prospect, and one of the main motivations in developing the ideas in this chapter was to pursue it.

II. Life-History Analysis and the Optimality Approach

Life-history analysis is the method of demonstrating quantitatively why some patterns of resource allocation are favored by natural selection in some environments and other patterns are favored in other environments. The form of the analysis is shown in Fig. 1a. The stippled set shows the possible ways in which resources can be allocated between two life-history characters (these might be somatic growth and reproduction, for example). Allocations around the star in Fig. 1a represent allocation of resources mostly to life-history character 2; those around the star in Fig. 1b represent allocation of resources mostly to life-history character 1. In the simplest case resource allocation is genetically controlled, and the stippled points represent the species' allelic possibilities for resource allocation. These

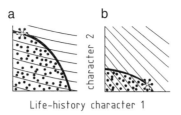

Life-history character 1

Figure 1   Methods of life-history analysis, where (a) and (b) represent different environments. Thick lines represent trade-off curves, dots represent genetic possibilities, and thin lines are fitness contours. Fitness increases to top right in both graphs. Stars represent optimal strategies/evolutionary outcomes.

genetic possibilities are subject to the energy allocation trade-off and this is often a straight line (see examples later in this chapter). However, it is not necessarily a straight line, and indeed is unlikely to be when one of the life-history characters represents survival. For this reason the trade-off is depicted as a curve in Fig. 1. Note that the trade-off curve represents the boundary of the genetic options set.

Superimposed on the graphs in Fig. 1 are *fitness contours* (thin lines). Each contour connects alleles with the same value of fitness. Fitness here represents the rate at which an allele spreads (or declines) in a population in a given environment. Provided only that the life history can be fully described, fitness is specified by the Euler–Lotka equation or a generalization of it, for species with nonmodular or modular growth, respectively (Sibly and Antonovics, 1992, Sibly and Curnow, 1993; Caswell, 1989). The methods for calculating fitness are well established, and this side of the analysis does not present major problems.

Both the shape of the trade-off curves and the shapes of the fitness contours may change between environments (Fig. 1). The evolutionary outcomes (i.e., the optimal energy allocations that have highest fitness, starred in Fig. 1) may therefore be different in the two environments, as shown in Fig. 1 for two isolated populations: in Fig. 1a the evolutionary outcome allocates most resources to life-history character 2, and in Fig. 1b to life-history character 1, as shown by the positions of the stars. While the analysis is straightforward if the populations are isolated, complications arise if there is gene flow between environments, since the fitness of an allele studied in one environment then depends at least to some extent on how it performs in the other environment (Houston and McNamara, 1992; Kawecki and Stearns, 1993; Sibly, 1995).

Since the shapes of the fitness contours are relatively easy to determine, the principal obstacle to progress in understanding why allocation patterns differ between environments has been in determining the shapes of the trade-off curves. Most available methods are cumbersome and time-consuming to use and generally yield only rather imprecise information about the shape of the trade-off curve (Reznick, 1985; Parker and Maynard Smith, 1990; Partridge and Sibly, 1991; Stearns, 1992; Roff, 1992). This is why if insights could be gained from a study of allocation constraints, they would be especially valuable.

III. Costs of Lignification in Early Seedling Growth

Early growth of seedlings has been studied with the aim of characterizing and classifying the types of reproductive strategies used by plants (Grime, 1979). It has long been known that woody plants grow more slowly than herbaceous plants, and it used to be thought that this was because their net photosynthetic rates were lower (Jarvis and Jarvis, 1964). Recent work, however, has shown that rates of photosynthesis and assimilation per unit leaf area are similar for woody and nonwoody angiosperms (Cannell, 1989). Here we make some quantitative predictions about the form of growth using the principles of energy allocation, and indicate in a preliminary way how these might be tested.

Suppose that in an "optimal growing environment" the energy (or strictly power) available to the seedling depends on leaf area. We shall suppose for simplicity that leaf area depends on leaf mass (more sophisticated analyses might use mass$^{2/3}$, or the mass of the part of the leaf that is involved in photosynthesis). Denoting leaf mass as m, we thus have

$$\text{Energy available} = A \times m, \tag{1}$$

where A is a constant of proportionality. Ignoring for the moment the costs of maintenance, we now suppose that the energy available is allocated either to leaf growth or to lignification, as shown in Fig. 2a. A fraction u of energy is allocated to leaf growth, and the remainder, $1 - u$, to lignification. Leaf growth is measured by dm/dt, and this depends both on the allocatable energy and on the fraction allocated to leaf growth. Thus, with suitably chosen units,

$$dm/dt = uAm, \tag{2}$$

which gives the exponential growth of leaf mass,

$$m = m_0 \, e^{uAt}, \tag{3}$$

where m_0 denotes leaf mass at time 0. It turns out that total plant mass, m_T, also grows exponentially, as will be seen [Eq. (8)]. If m_2 represents the mass of lignin, then

a

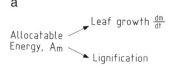

b

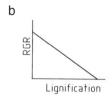

Figure 2   (a) Energy allocation diagram for the growth of seedlings that allocate available energy, *Am,* either to leaf growth or to lignification. (b) Predicted trade-off between relative growth rate (RGR) and lignification [Eq. (12)]. RGR is defined in Eq. (9).

$$m_T = m + m_2, \tag{4}$$

and

$$dm_2/dt = (1 - u)Am\eta_2, \tag{5}$$

where η_2 represents the conversion efficiency with which energy is converted to lignin mass. Equation (5) shows that lignin mass grows exponentially, since, inserting Eq. (3) into Eq. (5), we have

$$dm_2/dt = (1 - u)A\eta_2 m_0 e^{uAt}, \tag{6}$$

which can be integrated to give

$$m_2 = \frac{\eta_2(1 - u)}{u} m_0\, e^{uAt}. \tag{7}$$

Hence, combining Eqs. (3), (4), and (7), we obtain

$$m_T = m_0\, e^{uAt}\left[\frac{u + \eta_2(1 - u)}{u}\right]. \tag{8}$$

Equation (8) shows that total plant mass, m_T, grows exponentially. Relative growth rate (RGR), measured as $1/m_T\, dm_T/dt$, is given by

$$\text{RGR} = 1/m_T\, dm_T/dt = uA. \tag{9}$$

Thus, RGR is proportional to the fraction u of resources that are allocated to leaf growth.

Note that this argument does not depend on η_2, the efficiency with which energy is converted to lignin mass expressed as a fraction of the efficiency with which energy is converted to leaf mass. For some purposes, however, an estimate of η_2 will be required. The energy cost of producing a piece of wood or leaf may be guessed in a comparative way by looking at the energy required for biosynthesis. The production of 1 g of leaf, herbaceous stem, and woody stem takes about 0.23, 0.28, and 0.45 g of CO_2, respectively (Penning de Vries *et al.*, 1974; Cannell, 1989). However, the purely chemical approach does not include the energy required for the generation of shape. The total energy required for producing a structure also includes the energy required to make surfaces (internal or external) and the energy required to establish hydrophobic materials (e.g., lignin) in a hydrophilic environment such as the cellulose cell wall. No data are currently available on these energy costs.

It follows from Eqs. (2), (6), (7), and (9) that

$$RGR = 1/m_T \, dm_T/dt = 1/m \, dm/dt = 1/m_2 \, dm_2/dt. \qquad (10)$$

Equation (10) shows that the relative growth rates of m, m_2, and m_T are all equal; that is, the relative growth rates of the plant components are all equal, and are equal to the relative growth rate of the whole plant. In contrast, Hunt and Bazzaz (1980) suggested that the relative growth rate of the whole plant is equal to the algebraic sum of the relative growth rates of the plant's components.

For the purposes of what follows it is useful to rewrite Eq. (9) in terms of the fraction of energy allocated to lignification, which we now write as u_2, so that

$$u_2 = 1 - u \qquad (11)$$

and

$$RGR = A - Au_2. \qquad (12)$$

Thus, RGR is negatively proportional to the fraction u_2 of resources allocated to lignification. This prediction is illustrated in Fig. 2b.

Support for the prediction is shown in Fig. 3, where data on RGR_{max} from Grime and Hunt (1975) are plotted against a crude index of lignification assessed on a five-point scale. Point 0 on the scale represents no allocation to lignification, and point 5 would correspond to allocation of all resources to lignification (in which case there would be no scope for growth, and RGR would necessarily be zero). The regression equation fitted to the data of Fig. 3 is

$$RGR = 1.44 - 0.20 \times \text{lignification index,}$$

where RGR is in units of week^{-1} ($r^2 = 0.30$, slope not equal to zero, $p < 0.001$). The implication of this equation is that in the absence of lignification RGR is 1.44 week^{-1}, on average. For each one-point increase in the lignifica-

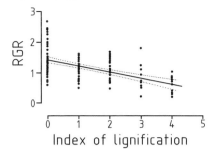

Figure 3   Relative growth rate (RGR, weeks^{-1}) plotted against an index of lignification, testing the prediction in Fig. 2b. See text for further details.

tion index, RGR decreases by 0.20 week^{-1}. This gives a measure of the energy cost of lignification, in units of "points lignification per unit decrease in RGR." Here, energy cost (strictly, power cost) is measured in units of lost RGR. In other words, energy cost is measured in terms of the growth rate that would have resulted if the energy had been allocated to growth. The regression equation also shows that RGR would be 0.44 week^{-1} at point 5 on the index of lignification, somewhat above the value (zero) predicted from Fig. 2b.

Further support for the prediction is provided by regression analysis of data on 14 grass species tabulated in Van Arendonk and Poorter (1994). The regression equation is

$$RGR = 0.305 - 0.0225 \times (\text{lignin} + \text{hemicellulose}),$$

where RGR is in units of day^{-1} and lignin and hemicellulose in units of g m^{-2} of leaf area ($r^2 = 0.68$, $F_{1,12} = 25.1$, $p < 0.001$). The implication of this equation is that for each additional gram of lignin plus hemicellulose in leaves of area 1 m^2, RGR decreases by 0.0225 day^{-1}. Observed values of RGR varied between 0.11 and 0.27 days^{-1} and those of lignin plus hemicellulose between 1.93 and 8.44 g m^{-2}.

The allocation diagram of Fig. 2a can be extended as in Fig. 4 to include all the types of defenses that are used by seedlings. Thus, u_3 represents the

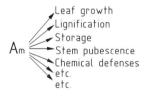

Figure 4   Extension of the energy allocation diagram of Fig. 2a to include the other types of defenses used by some species.

fraction of energy allocated to storage, u_4 the fraction allocated to stems, and so on. It turns out that the predicted growth of the whole plant is still exponential (Appendix 1) and that now

$$\mathrm{RGR} = A - Au_2 - Au_3 - \ldots - Au_k. \qquad (13)$$

Note that Eq. (13) is a simple generalization of Eq. (12), extending the two-way allocation diagram of Fig. 2a to the multiway allocation model of Fig. 4.

It would be interesting to test this allocation model using data from a variety of species, for example, by extending Fig. 3 to include the other processes to which seedlings allocate energy. This would involve a multiple regression of $\mathrm{RGR}_{\max}$ on %lignification, %storage, %stem, etc. As in Fig. 3 the regression coefficients can be interpreted as measures of the energy costs of lignification, storage, stem construction, etc., measured in units of lost RGR.

Note that u, the allocation to leaf growth, does not feature in Eq. (12), so the variables $u_2, \ldots, u_k$ represent $k - 1$ independent variables. However, correlations between these variables, if they exist, could cause problems in interpreting the output of the regression. If the model were successful then the residual variation unaccounted for by the model would be small.

The success of a model always depends on the assumptions on which it is based. The key assumptions here are as follows.

A. Assumptions

1. Seedling growth is measured under identical "optimal conditions" so that no resource other than incipient light is limiting.

2. The leaves of different species of seedlings are equally effective at converting solar radient energy into allocatable energy.

3. Allocatable energy is proportional to leaf mass.

4. Seedling growth is measured over a sufficiently short period of time that resource allocations (u_i) do not change with time and can be considered constant. Ontogenetic changes, though common in nature, are not considered here.

5. Costs of maintenance have been tacitly assumed to be the same in woody and nonwoody plants.

Provided these assumptions are reasonable, the results of the multiple regression can be used to calculate the energy costs of allocation to each process. This is most easily seen from Figs. 2b and 3 by noting that the slope of the line, which was obtained by regression, gives a measure of the energy costs of lignification.

IV. Benefits of Lignification in the Horizontal Branches of Trees

So far we have considered the energy costs of lignification, but we have not discussed its benefits. One of the benefits of lignification is in resisting gravity. This is most straightforward to analyze in the horizontal branches of trees. We therefore begin by considering how a horizontal branch resists the force of gravity. Our treatment builds on the earlier analyses of Alexander (1971), King and Loucks (1978), Cannell *et al.* (1988), Niklas (1992), Wainwright *et al.* (1976), and Givinish (1986).

At any position along the length of a horizontal branch (Fig. 5) three types of forces are acting: tension forces act horizontally in the branch's upper layers, compression forces act horizontally in its lower layers, and shear forces act vertically. The tension and compression forces together resist the leverage applied by the outer part of the branch. This leverage is measured as a *bending moment* and is exactly balanced by the bending moment supplied by the tension and compression forces acting together. These bending moments are more important and harder to resist than the shear forces, which we do not consider further here. The distribution of tensile and compressive forces within a tree varies between angiosperms (which can generate internal stresses by laying down pretensioned wood fibers) and gymnosperms (which do the job by growing wood in compression), so we are making some simplifying assumptions (Fournier *et al.*, 1994).

It follows from the above that

Prediction 1: If gravity is the main force acting on the branch then the major forces within it are tension in the upper layers and compression in the lower layers, and the structure of the wood should differ between the layers to resist these forces.

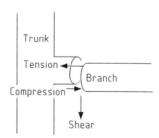

Figure 5   Schematic depiction of forces acting within the horizontal branches of trees.

The optimal cross section of branches or beams to withstand bending moments imposed by gravity is well known to be an I beam (Fig. 6a), because tension and compression forces acting furthest from the center contribute most to the bending moment (see, e.g., Wainwright *et al.*, 1976, p. 257). Clearly, however, branches are not I beams. One reason for this lies in the developmental constraint that larger branches grow from smaller branches, and if material cannot be reallocated then the cross sections of large branches must contain the cross sections of their smaller younger forms within them, as in Fig. 6b–d. This constraint gives the following:

Prediction 2: The cross section should be a debased I beam as in Fig. 6b, or perhaps a rectangular beam as in Fig. 6c.

However, even a rectangular beam is 18% better at resisting gravitational bending moments than a circular branch using the same amount of material. That debased I beams are achievable is shown by a remarkable photograph in Wood (1995) which resembles Fig. 6b without the corners.

Finally, we need to consider how the branch should taper along its length. A length of undivided branch without side branches is subject to gravitational forces on the wood within it and to an end load (and perhaps bending moments) imposed by the structures at its end. As a simplification we restrict attention to the end load and the weight of material within the branch (Fig. 7a).

Prediction 3: The optimal taper of a weightless branch carrying an end load follows a cube-root funtion, as in Fig. 7b. At the other extreme, i.e., a weighty branch with no end load, the optimal taper follows a quadratic function, as in Fig. 7c.

In reality branches are much closer to Fig. 7b than to Fig. 7c. In the few cases where the taper has been measured it is not as extreme as a cube-root function ($x^{1/3}$), but the exponent is nevertheless less than 1 (e.g., $x^{0.65-0.90}$; Burk *et al.*, 1983).

Putting together Predictions 1 to 3, it is apparent that branches are poorly designed if the main forces on them are those due to gravity: they

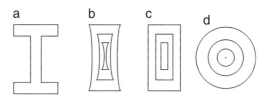

Figure 6 Cross sections of beams or branches discussed in the text.

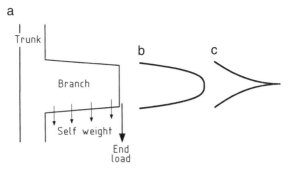

Figure 7 (a) Branch subject to gravitational forces on the wood within it and to an end load. (b) Optimal taper for a weightless branch carrying end load P. The equation is $r = (4Px/\pi\sigma)^{1/3}$ (Wainwright *et al.*, 1976, p. 249), where σ is the greatest horizontal stress the wood can sustain. (c) Optimal taper for a weighty branch with no end load. The equation is $r = 2\rho x^2/15\sigma$ where ρ is the density of the wood (Appendix 2).

are markedly suboptimal in several respects, and demonstrably superior forms can readily be found. Thus, the top layers should resist tensile forces and the bottom layers compression, the branch cross section should resemble a debased I beam, with the longitudinal taper ideally being closer to a quadratic form as in Fig. 7c. There seems no reason why these forms should not be realized in practice, and the fact that branches do not adopt such forms argues strongly that the starting premise is wrong, that gravitational forces are not the most important forces acting on horizontal branches. Probably end loadings imposed by wind action on leaves are much more important.

Such wind loading on leaves could presumably come from any direction, in which case a circular cross section (Fig. 6d) would be optimal, and top and bottom layers of the branch would both have to resist both tension and compression forces, though maximum stresses would still occur in the surface layers of the branch.

Of course branch design also has functions other than structural support, and these should ideally be taken into account in the optimality analysis (Farnsworth and Niklas, 1995). Transport of water and nutrients are obvious examples, and these may affect branch design because water flows best when wood is more porous.

Our analysis shows that branches are not optimally or even well designed to withstand gravitational forces, but they do seem well designed, perhaps optimally designed to withstand wind-imposed end loads. Branches in more exposed places should be thicker, and branch diameters should be related to the wind resistance of the leaves they carry. Furthermore, the cross-sectional form of the branches reflects the direction of the strongest winds they experienced, and material tends to be added according to where the greatest stresses have been experienced (Fournier *et al.*, 1994). Further useful discussion of these issues can be found in King and Loucks (1978) and in papers in Coutts and Grace (1995).

V. General Discussion

The principles of resource allocation have been used as a basis from which to analyze resource allocation in seedling growth and in the design of branches. The principal conclusions are as follows. In analyzing seedling growth, multiple regression should be used to investigate the dependence of relative growth rate (RGR) on the various defense systems employed by seedlings (examples are given in Fig. 4). The regression coefficients can be readily interpreted in terms of the energy costs of the various defense systems. The main conclusion as to the design of branches is that branches are not well designed if they are only to withstand gravitational forces. However, they do seem well designed—perhaps optimally designed—to withstand wind-imposed end loads.

Although these analyses offer some fresh insights, they have not provided the hoped-for information as to the form and shape of life-history trade-offs. We now discuss the reason for this. In both examples the problems stem from the fact that we do not know the the detailed survivorship implications of defense allocation.

The first step in the life-history analysis is to establish whether the "allocation trade-off" (e.g., Fig. 2b) is indeed linear. Linearity follows logically if resources allocated to one function are not available to others. However, more complicated cases are possible if structures have more than one function. For instance, spike leaves may both generate allocatable energy and provide defense. Figure 3 provides some evidence that the allocation trade-off may indeed be linear, but the best evidence of linearity would be a good fit to data using the multiple regression model suggested for the analysis of Fig. 4.

In this way resource allocation to growth, reproduction, and defense may be shown to result in a linear trade-off as in Figs. 2 and 4. Alternatively, if the allocation trade-off is not linear it may nevertheless be possible to ascertain its shape. In the life-history analysis of Fig. 1, however, the allocation trade-off has to be converted to life-history variables. Suitable life-history variables would be (1) ages at which reproduction occurs, (2) fecundities then achieved, and (3) survivorships between reproductive events. Juvenile growth rate affects the age at which reproduction occurs, and this and fecundity are readily measurable. More problematical is ascertaining how resource spending on defense translates into improved survivorship.

The needed analysis can be illustrated by a simple graphical example (Fig. 8). Suppose the allocation trade-off, between, for example, reproduction and defense, is known and is found to be linear (Fig. 8a). Suppose also that the survivorship implications of spending on defense are known (Fig. 8b). Now the defense variable can be eliminated between Fig. 8a and Fig. 8b, and the result is the required relationship between reproduction and survivorship (Fig. 8c). Note that this is in the same form as Fig. 1. Such a curve gives explicit form to the trade-off between economy and safety proposed by Chazdon (1986).

In terms of the two examples discussed in this chapter, experiments would be required to investigate the dependence of survivorship on different types of spending on defense. In the case of seedlings these would ideally be field experiments that would compare the survivorships of seedlings more or less well defended with each of the defenses illustrated in Fig. 4. In the case of branches it is necessary to (1) experimentally manipulate branch design, (2) investigate the breaking loads that can be withstood by the different designs, perhaps in a wind tunnel, and (3) relate these to the incidence of winds of different speeds and directions in the field.

VI. Conclusions

Consideration of resource allocation during seedling growth suggests the existence of a negative trade-off between lignification and growth rate. Some evidence for such a trade-off is presented. The analysis can be generalized using multiple regression; the regression coefficients then have a direct interpretation in terms of the energy costs of individual defense systems. Resource allocation principles are used more loosely in the second half of the chapter to analyze the three-dimensional structure of branches.

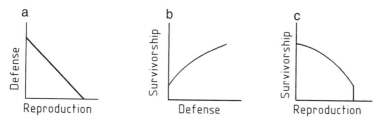

Figure 8   Construction of a life-history trade-off curve. (a) Linear trade-off between resources allocated to reproduction and those allocated to defense. (b) Curve specifying how defense allocation improves survivorship, which is unknown in our analyses. (c) Eliminating the defense variable between (a) and (b) allows construction of the needed life-history trade-off relating reproduction to survivorship.

Consideration of (1) the shape of the cross section, (2) the material used in the cross section, and (3) the form of the longitudinal taper all suggest that branch design is poor if the main forces on the branch are gravitational. Branch design may, however, be optimal if the main forces are those imposed by wind loadings. This suggests that branch design provides a three-dimensional record of the directions and speeds of the major winds to which the branch has been exposed.

Finally some problems with optimality analyses are reviewed, and it is noted that allocations between growth and reproduction translate relatively simply and directly into life-history trade-offs. By contrast trade-offs involving mortality are harder to characterize, because of the difficulty of measuring the survivorship benefits of allocating resources to defense.

VII. Appendix 1

Plant Growth Is Exponential for the Multiway Allocation of Energy Shown in Figure 4

The proof is an extension of the two-way allocation treated in Eqs. (1)—(12). Let m_i represent the mass of structure i so that total mass m_T is given by

$$m_T = m + m_2 + m_3 + \ldots + m_k \tag{A1}$$

[cf. Eq. (4)]. We assume that the rate of growth of structure i, dm_i/dt, depends on the energy allocated to it; thus,

$$dm_i/dt = u_i Am\eta_i \tag{A2}$$

[cf. Eq. (5)], where η_i represents the "conversion efficiency" with which energy is converted to mass of structure i. Remember that m in Eq. (A2) represents leaf mass. Using the methods of Eqs. (6) and (7),

$$m_i = \frac{\eta_i u_i}{u} m_0 e^{uAt}. \tag{A3}$$

Hence, combining Eqs. (3), (A1), and (A3), we obtain

$$m_T = m_0 e^{uAt} \left(\frac{u + \Sigma_2^k \eta_i u_i}{u} \right), \tag{A4}$$

which shows that total plant mass, m_T, grows exponentially. Relative growth rate (RGR), measured as $1/m_T \, dm_T/dt$, is given by Eq. (9), as before, and since

$$u = 1 - u_2 - u_3 - u_4 - \ldots - u_k, \tag{A5}$$

Eq. (9) can also be written

$$RGR = A - Au_2 - Au_3 - Au_4 - \ldots - Au_k,$$

which is Eq. (13).

VIII. Appendix 2

The Optimal Taper for a Weighty Branch with No End Load Follows a Quadratic Function, as in Figure 7c

Suppose the radius r of a branch of circular cross section varies with the distance from its tip, x, according to $r = r(x)$. We seek the form of $r(x)$ that keeps the maximum stress σ the same in all sections.

Let distance x from the branch tip be labeled C. The bending moment sustained at C is

$$\frac{\sigma \pi \, [r(x)]^3}{4} \tag{A6}$$

(see, e.g., Wainwright *et al.*, 1976). This counteracts the bending moment imposed by the outer branch, which is given by

$$\int_0^x (x - y)\rho\pi[r(y)]^2 \, dy, \tag{A7}$$

where ρ is the density of the wood.

Setting expression (A6) equal to expression (A7) gives

$$\frac{\sigma\pi[r(x)]^3}{4} = \int_0^x (x - y)\rho\pi[r(y)]^2 \, dy. \tag{A8}$$

We seek a function $r = r(x)$ that solves Eq. (A8). Suppose the solution has the form $r = zx^2$ where z is a constant to be found. Then

$$\frac{\sigma\pi z^3 x^6}{4} = \frac{\rho\pi z^2 x^6}{5} - \frac{\rho\pi z^2 x^6}{6} = \frac{\rho\pi z^2 x^6}{30}, \tag{A9}$$

giving $z = 2\rho/15\sigma$. Hence,

$$r = \frac{2\rho}{15\sigma} \, x^2 \tag{A10}$$

solves Eq. (A8) and so gives the optimal taper.

Acknowledgments

We thank J. Barnett, Dr. P. G. Sibly, and particularly Dr. R. Hunt and an anonymous referee for comments and suggestions.

References

Alexander, R. M. (1971). "Size and Shape." Arnold, London.

Bartareau, T. (1995). Pollination limitation, costs of capsule production and the capsule-to-flower ratio in *Dendrobium monophyllum* F. Muell. (Orchidaceae). *Aust. J. Ecol.* **20,** 257–265.

Burk, T. E., Nelson, N. D., and Isebrands, J.G., (1983). Crown architecture of short-rotation, intensively cultured *Populus.* III. A model of first-order branch architecture. *Can. J. For. Res.* **13,** 1107–1116.

Cannell, M. G. R. (1989). Physiological basis of wood production: A review. *Scand. J. For. Res.* **4,** 459–490.

Cannell, M. G. R., Morgan, J., and Murray, M. B. (1988) Diameters and dry weights of tree shoots: Effects of Young's modulus, taper, deflection and angle. *Tree Physiol.* **4,** 219–231.

Caswell, H. (1989). "Matrix Population Models." Sinauer, Sunderland, Massachusetts.

Chazdon, R. L. (1986). The costs of leaf support in understory palms: Economy versus safety. *Am. Nat.* **127,** 9–30.

Coutts, M. P., and Grace, J. (eds.) (1995). "Wind and Trees." Cambridge Univ. Press, Cambridge.

Fagerstrom, T., Larsson, S., and Tenow, O. (1987). On optimal defense in plants. *Funct. Ecol.* **1,** 73–82.

Farnsworth, K. D., and Niklas, K. J. (1995). Theories of optimization, form and function in branching architecture in plants. *Funct. Ecol.* **9,** 355–363.

Fournier, M., Bailleres, H., and Chanson, B. (1994). Tree biomechanics: Growth, cumulative prestresses, and reorientations. *Biomimetics* **2,** 229–251.

Fox, G. A. (1992). Annual plant life histories and the paradigm of resource allocation. *Evol. Ecol.* **6,** 482–499.

Givinish, T. J. (ed.) (1986). "On the Economy of Plant Form and Function." Cambridge Univ. Press, Cambridge.

Grime, J. P. (1979). "Plant Strategies and Vegetation Processes." Wiley, Chichester.

Grime, J. P., and Hunt, R. (1975). Relative growth rate: Its range and adaptive significance in a local flora. *J. Ecol.* **63,** 393–422.

Hara, T., Kawano, S., and Nagai, Y. (1988). Optimal reproduction strategy of plants with special reference to the modes of reproductive resource allocation. *Plant Species Biol.* **3,** 43–60.

Houston, A. I., and McNamara, J. M. (1992). Phenotypic plasticity as a state dependent life-history decision. *Evol. Ecol.* **6,** 243–253.

Hunt, R., and Bazzaz, F. A. (1980). The biology of *Ambrosia trifida* L. V. Response to fertilizer, with growth analysis at the organismal and sub-organismal levels. *New Phytol.* **84,** 113–121.

Iwasa, A., and Cohen, D. (1989). Optimal growth schedule of a perennial plant. *Am. Nat.* **133,** 480–505.

Iwasa, Y. (1990). Evolution of the selfing rate and resource allocation models. *Plant Species Biol.* **5,** 19–30.

Jarvis, P. G., and Jarvis, M. S. (1964). Growth rates of woody plants. *Physiol. Plant.* **17,** 654–666.

Kawecki, T. J., and Stearns, S. C. (1993). The evolution of life histories in spatially heterogeneous environments: Optimal reaction norms revisited. *Evol. Ecol.* **7,** 155–174.

King, D., and Loucks, O. L. (1978). The theory of tree bole and branch form. *Radiat. Environ. Biophys.* **15,** 141–165.

McGinley, M. A., and Charnov, E. L. (1988). Multiple resources and the optimal balance between size and number of offspring. *Evol. Ecol.* **2,** 77–84.

Mazer, S. J. (1987). Maternal investment and male reproductive success in angiosperms: Parent offspring conflict or sexual selection. *Biol. J. Linn. Soc.* **30,** 115–134.

Niklas, K. J. (1992). "Plant Biomechanics." Univ. of Chicago Press, Chicago.

Parker, G. A., and Maynard Smith, J. (1990). Optimality in evolutionary biology. *Nature (London)* **348**, 27–33.

Partridge, L., and Sibly, R. (1991). Constraints on the evolution of life histories. *Philos. Trans. R. Soc. London B* **332**, 3–13.

Penning de Vries, F. W. T., Brunsting, A. H. M., and van Laar, H. H. (1974). Products, requirements and efficiency of biosynthesis: A quantitative approach. *J. Theor. Biol.* **45**, 339–377.

Perrin, N., and Sibly, R. M. (1993). Dynamic models of energy allocation and investment. *Annu. Rev. Ecol. Syst.* **24**, 379–410.

Pugliese, A. (1987). Optimal resource allocation and optimal size in perennial herbs. *J. Theor. Biol.* **126**, 33–50.

Pugliese, A. (1988). Optimal resource allocation in perennial plants: A continuous-time model. *Theoretical Population Biology* **34**, 215–247.

Ravishankar, K. V., Shaanker, R. U., and Ganashai, K. N. (1995). War of hormones over resource allocation to seeds: Strategies and counter-strategies of offspring and maternal parent. *J. Biosci.* **20**, 89–103.

Reznick, D. (1985). Costs of reproduction: An evaluation of the empirical evidence. *Oikos* **44**, 257–267.

Roff, D. A. (1992). "The Evolution of Life Histories: Theory and Analysis." Chapman & Hall, London.

Sakai, S. (1995a). Optimal resources allocation to vegetative and sexual reproduction of a plant growing in a spatially varying environment. *J. Theor. Biol.* **175**, 271–282.

Sakai, S. (1995b). A model for seed size variation among plants. *Evol. Ecol.* **9**, 495–507.

Sakai, S. (1995c). Evolutionary stable selfing rates of hermaphroditic plants in competing and delayed selfing modes with allocation to attractive structures. *Evolution* **49**, 557–564.

Schoen, D. J., and Ashman, T.-L. (1995). The evolution of floral longevity: Resource allocation to maintenance versus construction of repeated parts in modular organisms. *Evolution* **49**, 131–139.

Sibly, R. M. (1995). Life-history evolution in spatially-heterogeneous environments, with and without phenotypic plasticity. *Evol. Ecol.* **9**, 262–267.

Sibly, R. M., and Antonovics, J. (1992). Life-history evolution. *In* "Genes in Ecology" (R. J. Berry, T. J. Crawford, and G. M. Hewitt, eds.), pp. 87–122. Blackwell, Oxford.

Sibly, R. M., and Calow, P. (1986). "Physiological Ecology of Animals." Blackwell, Oxford.

Sibly, R. M., and Curnow, R. N. (1993). An allelocentric view of life-history evolution. *J. Theor. Biol.* **160**, 533–546.

Stearns, S. C. (1992). "The Evolution of Life Histories." Oxford Univ. Press, Oxford.

Van Arendonk, J. J. C. M., and Poorter, H. (1994). The chemical composition and anatomical structure of leaves of grass species differing in relative growth rate. *Plant Cell Environ.* **17**, 963–970.

Wainwright, S. A., Biggs, W. D., Currey, J. D., and Gosline, J. M. (1976). "Mechanical Design in Organisms." Arnold, London.

Wood, C. J. (1995). Understanding wood forces on trees. *In* "Wind and Trees." (M. P. Coutts and J. Grace, eds.), pp. 133–164. Cambridge Univ. Press, Cambridge.

7

Resource Allocation Patterns in Clonal Herbs and Their Consequences for Growth

Michael J. Hutchings

I. Introduction

Plants of most species are assemblages of repeated modules. Each module consists of a segment of stem that includes a node, at which organs for resource acquisition (leaves and in some species roots) may be produced, and an axillary bud which, if it grows, can either generate more modules or produce a structure such as an inflorescence, which terminates the growth of that axis. The number and spatial configuration of modules are highly variable between different plants in most species. Modular plants grow by generating new modules, and consecutive modules on any axis are separated by stem internodes. Thus, the modules of such plants are dispersed in space and time (Mogie and Hutchings, 1990). As resource availability varies in most habitats in space and time (e.g., Grime, 1994: Bell and Lechowicz, 1994), often at small scales, different modules of the same plant will frequently occupy sites that differ in quality.

Studies of plasticity in resource allocation in response to growing conditions are abundant in the ecological literature (e.g., Harper and Ogden, 1970; van Andel and Vera, 1977; Waite and Hutchings, 1982; Bazzaz and Reekie, 1985; Clauss and Aarssen, 1994). Modular species have usually been studied, and analyses have usually been carried out at the whole plant level. Although this usually seems appropriate, especially when the species has a single rooting site, it may not always be, because plants often respond to their growing conditions at smaller structural scales. Theory suggests that resources can be acquired most efficiently from localities where concentration is high (Bloom *et al.,* 1985). Essential resources often show negative spatial covariance (Schlesinger *et al.,* 1990; Friedman and Alpert 1991; Stuefer and Hutchings, 1994), so that, if plants behaved according to theory, they would exhibit plasticity promoting acquisition of locally abundant resources from different microsites within their habitat. Thus, under heterogeneous conditions, modules should show localized plasticity that enhances the acquisition of locally abundant resources. If each module can produce both roots and leaves, and therefore acquire *all* essential resources for growth, it should be possible to observe localized plastic responses to local resource supply in heterogeneous habitats. There might then be a division of labor in the resource acquisition activities of different modules.

Under favorable conditions, many modular species produce roots as well as leaves at their nodes (Mogie and Hutchings, 1990), and in many modular species with plagiotropic stems (i.e., stems growing in the horizontal plane), root production from nodes is normal, conferring clonality. In these species each module can therefore acquire all essential resources and potentially survive if separated from the parent plant. Unless they fragment, clonal species thus acquire resources at many "feeding sites" (Bell 1984) or "mouths" (Watson, 1984) that are separated by internodes or "spacers" (Bell 1984). The term ramet is commonly used as a synonym for module in these species (Jackson *et al.,* 1985). The ramets of individual plants may be distributed widely in time and space in some of these species; genetical individuals may survive for decades or centuries, distributing huge numbers of ramets over wide areas (Cook, 1983). The connected ramets of single clones therefore commonly experience a variety of growing conditions, and may be expected to display plasticity in resource allocation on a finer structural scale than that of the whole plant.

This chapter reviews the effects of habitat conditions on allocation patterns in clonal plants, mostly using information from experiments carried out under controlled conditions. First, resource allocation patterns are compared in different homogeneous conditions. Second, the effects of exposing specific substructures of single clones to different growing conditions are described. Finally, comparisons are made of clone growth when a given quantity of nutrients is provided in different spatial configurations. Proportional allocation of biomass to different structures is emphasized, but other properties are considered, in particular the timing of growth of various structures and the ways in which they are constructed.

Earlier papers have discussed the relationship between allocation of resources to sexual and vegetative functions in clonal species (e.g., Ogden, 1974; Holler and Abrahamson, 1977; Armstrong, 1982; Loehle, 1987; Hartnett, 1990). This topic is not addressed here. Similarly, carbohydrate, water, and nutrient distribution patterns in clonal plants, which are highly dependent on conditions, plant structure, and vascular architecture, are not considered; these subjects have been reviewed by Marshall (1990).

II. Interpretation of Data on Resource Allocation Patterns

Plasticity in the resource allocation patterns of plants under different growing conditions is often assumed to be adaptive, and to reflect a tendency toward optimization of performance in any environment. However, resource partitioning also changes significantly during plant growth (Evans, 1972; Bradbury and Hofstra, 1976; Thompson and Stewart, 1981; Samson and Werk, 1986; Weiner, 1988; Clauss and Aarssen, 1994; Coleman *et al.*, 1994; de Kroon *et al.* 1994). Growing conditions affect plant growth, and thus the effects of growing conditions on resource partitioning interact with those of ontogeny. Consequently, differences in resource partitioning cannot be ascribed solely either to growing conditions or to ontogenetic state unless certain precautions are met. Changes in resource partitioning by plants growing under a given set of conditions, or experiencing seasonal changes, can easily be analyzed by periodic harvesting (Fig. 1), but unambiguous interpretation of the effects of different growing conditions on partitioning is possible only when plants are compared at the same size or developmental stage, rather than at the same age. Until relatively recently this approach has rarely been adopted in studies of either clonal or nonclonal species, and therefore many comparisons of resource partitioning under different growing conditions are difficult to interpret, as examples quoted by Coleman *et al.* (1994) demonstrate.

Growing conditions have dramatic effects on branching and other aspects of clone morphology (e.g., de Kroon and Hutchings, 1995), making size and weight difficult to standardize between treatments. Thus, it is often easier to compare clones at the same developmental stage, rather than at the same size. The most convenient measure of developmental stage is the plastochron index (PI), a plastochron being the time that elapses between adjacent modules or ramets on a given axis achieving the same developmental state. The PI can reflect both the time taken by clones to achieve a given status and the number of primary ramets that they have produced (Birch and Hutchings, 1992). As development of higher order axes is so dependent on growing conditions, the primary (lowest order) axis is normally used for assessing developmental state. Thus, the number of plastochrons for which the primary axis has grown is often standardized across treatments. The plastochron for primary axes is very stable in *Glechoma hederacea* over a wide range of growing conditions, so that it can be used both as an index of clone development and for setting the duration of

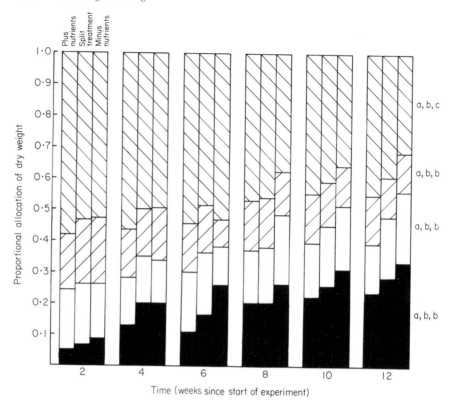

Figure 1   Proportional allocation of total biomass by clones of *Glechoma hederacea*, in three nutrient treatments, to leaves (◫), petioles (▨), stolons (□), and roots (■), throughout the course of a 12-week experiment. Significant differences between the proportions of biomass allocated to different components at the final harvest are indicated by different letters in the following order: high nutrient level, split treatment, low nutrient level. From Slade and Hutchings (1987a).

experiments. This advantage is made use of in several of the studies described below.

III. Resource Allocation Patterns of Clonal Species in Homogeneous Growing Conditions

A. Studies on Clonal Species under Field Conditions

Comparisons of resource partitioning in contrasting natural habitats are often unsatisfactory because plant material from different sources is hard to standardize, and because habitats (apart from being intrinsically heterogeneous) usually differ in several variables, making the cause of any difference hard to determine. There are few such comparisons for clonal plants. Lovett Doust (1981a) showed that resource partitioning differed in grassland and woodland clones of *Ranunculus repens*, both at the whole clone and daughter ramet level. Grassland clones allocated more of their biomass to leaves than woodland clones until July, after which they invested heavily in daughter ramet leaves while woodland clones invested more of their biomass in parent rosette leaves and stolons. Daughter ramets from both habitats had the same total dry weight in July. Those from grassland allocated more biomass to roots and less to stolons than those from woodland. Allocation to leaves did not differ significantly. d'Hertefeldt and Jónsdóttir (1994) compared resource partitioning in *Maianthemum bifolium* in sites differing in ground layer diversity and cover, and in tree canopy cover. More biomass was invested in leaves and less in rhizomes and roots where ground layer competition was high and tree cover low, although the differences were not significant. Rhizomes were shorter and had higher specific mass ($P < 0.001$) where there was more ground cover and a more open tree canopy.

Pitelka *et al.* (1985) showed that although the absolute amount of biomass in rhizomes was always higher in ramets of *Aster acuminatus* given high light late in the growing season, relative allocation showed no clear relationship with the temporal pattern of light supply. Allocation to reproduction increased with the amount of high intensity light received prior to flowering, but high light close to flowering time was most effective.

B. Studies of Resource Partitioning under Experimental Conditions

Experimental studies, as described below, provide a more complete picture of biomass partitioning in clonal species in response to environmental variables. First, some results are described from a unique study in which the response of a clonal species is documented across a complete range of resource supply from extreme deficiency to extreme toxicity.

1. Biomass Partitioning across a Resource Gradient   The literature on experiments that investigate resource allocation patterns in response to growing conditions is littered with contradictions. In one study, allocation to one function increases as the supply of some external variable is increased, whereas in another study allocation to the same function decreases. Such contradictions may reflect real differences in species' behavior, but they can also be caused by the levels of the external variables chosen for the experiment. We rarely know in advance how the behavior of a species alters with resource supply, the resource level at which allocation to any function reaches its maxima or minima, or the range of values over which allocation to the function fluctuates. However, de Kroon and Schieving (1991) have published a model attempting to predict the likely patterns of allocation of biomass to different structural organs as resource supply varies. Starting from empirical observations that branching intensity in clonal plants increases with increasing habitat quality, while internode length may decrease, a model was constructed allowing prediction of the optimal proportion of resources which clonal species should allocate to spacers (either stolons or rhizomes). The prediction is made that resource allocation to spacers should be greater at intermediate levels of resource supply than at high or low levels.

This model has been tested on *G. hederacea* clones using nutrient supply as the variable. Altogether 12 nutrient levels were used, increasing in a geometrical progression from extreme deficiency to extreme toxicity (1/64 strength to 32 times full strength Hoagland's solution). Clones were grown with each ramet rooted individually in pots of sand to which specified amounts of the appropriate nutrient solution were added (see Pelling, 1994, for full details). This many treatments in experiments on clonal plants could be a logistic nightmare, but the use of smoothing techniques to characterize the responses of the species across the gradient enables replication to be safely reduced without sacrificing insight into the nature and degree of response to environmental conditions (Austin and Austin, 1980): as replicates can be cloned from a single genotype, even small numbers of replicates give reliable results.

Secondary stolon production was highly volatile in response to nutrient level. When measured at a standard PI, it was seen to peak in the middle of the nutrient range and to fall rapidly with deficient and toxic nutrient levels (Fig. 2a), and, as primary ramets accumulated, the tendency for more stolon branches to develop at moderate nutrient levels grew more striking. The number of secondary ramets produced by the clone further emphasizes this volatility in branching (Fig. 2b). Although many aspects of clone growth achieved peak values at the same nutrient strength (standard strength Hoagland's solution), suggesting coordination of many aspects of clone growth, this cannot be true for biomass partitioning, because distribution of a finite resource between different structures must involve trade-offs. Allocation of biomass to leaves fell from 50% at the highest nutrient level to 25% at the lowest nutrient level (Fig. 3). In contrast, allocation to roots rose from less than 15% at the highest nutrient level to over 40% at the lowest. The proportion of biomass in petioles increased steadily as nutrient supply increased. Allocation to stolons was lowest in the middle of the nutrient gradient, contradicting the prediction of de Kroon and Schieving (1991). Whether this is due to ontogenetic differences between clones that have grown to different extents in the different nutrient supply treatments is not known, but this appears to be unlikely, given that the major difference

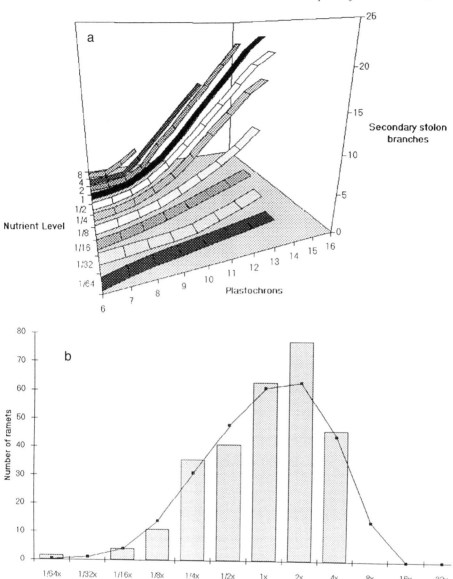

Figure 2   (a) Number of secondary stolons produced by clones of *Glechoma hederacea* when grown at different nutrient levels as a function of plastochron age of the primary stolon. (b) Secondary ramet production by clones of *Glechoma hederacea* when grown with different nutrient levels. Bars are measured values, and points are smoothed values (Pelling 1994).

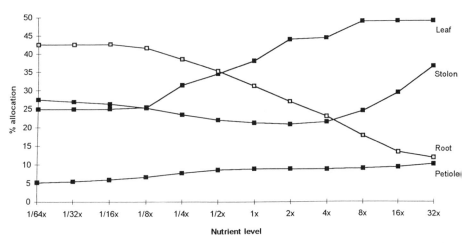

Figure 3   Proportional allocation of biomass to four component structures by clones of *Glechoma hederacea* when grown at different nutrient levels. All values are smoothed (Pelling 1994).

between clones in the different treatments is the extent of module proliferation.

2. Response to Nutrients   When the clonal graminoids *Ammophila arenaria*, *Elymus mollis* (Pavlik, 1983), and *E. repens* (Neuteboom and Cramer, 1985) and the vine *Calystegia sepium* (Klimeš and Klimešová 1994) were grown under high and low levels of nitrogen supply in homogeneous conditions, the root : shoot (R/S) ratio was higher when nitrogen supply was low. Regardless of nutrient supply, the growth of *C. sepium* was best when it experienced intraspecific shoot competition, and these larger plants allocated more of their biomass to runners under nutrient-poor than nutrient-rich conditions. This may increase the probability of finding uncontested patches of nutrients in a heterogeneous environment. *Elymus repens* rhizomes branched more under higher nitrogen levels, a response that may produce an increase in the density of feeding sites in localities where nutrients are abundant. [Hutchings and de Kroon (1994) have reviewed the detection of and response to local environmental quality, as reflected by the availability to plants of several variables including nutrient supply.]

Clones of *Glechoma hederacea* (Slade and Hutchings, 1987a; Fig. 1) and single ramets of *Fragaria chiloensis* (Alpert, 1991) also allocated more of their biomass to belowground parts under homogeneous nutrient-poor conditions. The difference was shown to persist in *G. hederacea* as the clones grew. The complex shifts and trade-offs in the patterns of allocation of biomass to different structures at different nutrient levels (Table I) are difficult to interpret, because of differences in growth in each treatment. Allocation to leaves fell by about one-third under low nutrient supply, and a significantly higher proportion of biomass was allocated to roots and stolons. As in *C. sepium,* the increased allocation to stolons may increase the probability of apices escaping from sites where nutrients are scarce and growing into a more favorable habitat. In *G. hederacea,* the increased allocation to stolons under low nutrient supply was accompanied by a reduction in weight per unit length of stolon and an increase in internode length. In contrast, single ramets of *F. chiloensis* allocated more of their biomass to stolons as nitrogen supply increased.

3. Response to Light Regime   Experiments on *Glechoma hederacea* showed that the proportion of biomass allocated to leaves and stolons was little changed by the availability of photosynthetically active radiation (PAR)—although the differences were significant—when nutrients were abundant, whereas the proportion of total biomass in petioles increased by 10% when PAR was reduced to 25% of ambient, and the proportion in roots decreased by 16% (Table II; Slade and Hutchings, 1987b). Comparisons of biomass partitioning under high and low PAR at low nutrient levels could not be made due to the poor growth of the species when given both low nutrients and low light. There were considerable differences in the leaves and stolons produced under different light PAR levels. Specific leaf area (SLA) was more than three times higher, stolon internodes were significantly longer, and dry weight per unit length of stolon was significantly less under low PAR (Table III). As in other clonal species (Solangaarachchi and Harper,

Table I   Proportional Allocation of Dry Weight and Root/Shoot (R/S) Ratio on Primary Structures of *Glechoma hederacea* to Different Organs after 12 Weeks of Growth in Experimental Conditions[a]

	Pure treatments		Split treatment	
	Nutrient poor	Nutrient rich	Nutrient poor	Nutrient rich
Allocation[b]				
Leaves	30.4 ± 1.6a	46.8 ± 1.4b	29.4 ± 1.5a	44.9 ± 1.5b
Petioles	10.5 ± 0.4a	10.8 ± 0.5a,b	9.6 ± 0.7a	9.4 ± 0.9a,c
Stolons	16.7 ± 1.1a	10.7 ± 0.4b	13.5 ± 0.4c	11.2 ± 0.4b,c
Roots	42.4 ± 2.2a	31.7 ± 1.5b	47.5 ± 2.4a	34.5 ± 2.2b
R/S ratio	0.74	0.46	0.90	0.53

[a] Slade and Hutchings (1987a).
[b] Data are mean percentages ± S.E. ($n = 6$). Within each line, values followed by the same letter are not significantly different at $P < 0.05$ or greater, after arcsine transformation.

Table II   Percentage Allocation of Biomass to
Four Structural Components, by *Glechoma
hederacea* Clones Grown under
Different Conditions[a]

	+L+N	+L−N	−L+N
Leaves	38 ± 1a	32 ± 1b	41 ± 1a
Petioles	13 ± 1a	10 ± 1b	23 ± 1c
Stolons	15 ± 1a	18 ± 1b	18 ± 1b
Roots	34 ± 1a	40 ± 1b	18 ± 1c

[a] In +L conditions plants received ambient radiation, whereas in −L conditions photon flux density was 25% of ambient. In +N conditions all ramets were watered periodically with a full-strength nutrient solution, whereas in −N conditions ramets received one-eighth strength nutrient solution. Further details can be found in Slade and Hutchings (1987c).

1987; Dong, 1993; de Kroon and Hutchings, 1995), petiole length was highly responsive in *G. hederacea*, increasing significantly as PAR fell (Price and Hutchings, 1996). The results for *G. hederacea* and for *Lamiastrum galeobdolon*, another member of the Lamiaceae, are strikingly similar. Both also show significant reductions in weight per unit length of petiole at low PAR (Dong, 1993; Price and Hutchings, 1996).

Plasticity in stolon morphology may promote ramet placement in the more favorable patches of a heterogeneous habitat. In contrast, longer-lived rhizomes may function mainly as storage organs. Stolons may therefore be more plastic than rhizomes in response to resource supply, but if storage falls when resources are scarce, allocation of biomass to rhizomes would be more responsive to resource supply than allocation to stolons. These hypotheses were tested in a factorial experiment with two light and two nutrient levels, on *Cynodon dactylon*, a clonal species that produces both stolons and rhizomes (Dong and de Kroon, 1994). Both variables affected

Table III   Disposition of Allocated Biomass within Clones of *Glechoma hederacea* under
Different Growing Conditions

	+L+N	+L−N	−L+N
Mean internode length (cm)	5.8 ± 0.2a	6.9 ± 0.1b	9.5 ± 0.3c
Dry weight per unit length of stolon (g cm^{-1} × 10^{-4})	35.4 ± 1.6a	35.7 ± 1.1a	12.4 ± 0.6b
Specific leaf area (cm^2 g^{-1})	143 ± 19a	134 ± 5a	461 ± 59b

stolon branching, but only nutrients affected rhizome branching. Stolon internodes were significantly longer under low PAR, but nutrients had little effect. Neither variable influenced rhizome lengths. Thus, stolons exhibited more morphological plasticity. Although plant size affected allocation patterns, there were also significant effects of light on allocation to stolons and of nutrients on rhizome allocation. The slopes of the regressions between stolon or rhizome mass and leaf + root mass were less steep at lower resource levels, indicating reduced allocation to both of these structures when resources were scarcer, but the effect on rhizomes was significantly stronger, supporting the hypothesis that resource shortage affects storage organs more than foraging organs.

These results were confirmed and extended in a study of the responses of three grass species to light supply (Dong and Pierdominici, 1995). One species (*Agrostis stolonifera*) was stoloniferous, one (*Holcus mollis*) rhizomatous, and the third (as above, *Cynodon dactylon*) produces both stolons and rhizomes. Biomass partitioning to rhizomes was lower at lower PAR, whereas partitioning to stolons was unaffected by light. In most cases rhizomes possessed more dormant buds than stolons. Whereas stolon internodes were longer under lower PAR, rhizome morphology did not respond significantly to light availability. These results again suggest that stolons primarily forage for light, a function they can perform efficiently because they are highly plastic. Rhizomes function as stores of carbohydrates but also of dormant meristems, and thus offer much potential for resuming growth if the plant is damaged (see also Eriksson 1985).

In another study (Huber, 1996), two stoloniferous *Potentilla* species, *P. anglica* and *P. reptans*, were grown under either ambient levels of PAR, neutral shading to 24% of ambient PAR, or shading to 24% of ambient PAR with a reduced red : far red (R/FR) ratio. As far as possible, plants were harvested at the same developmental stage, and divided into different organs. For both species, the R/S ratio decreased significantly under neutral shade and further still when the plants were grown under a low R/FR ratio (Fig. 4a). In both species the proportion of biomass allocated to petioles increased substantially under shading, whereas allocation to stolons fell slightly in *P. anglica* and considerably in *P. reptans*. Whereas the absolute weights of the main stem axes fell under shading in both species, petiole weights were significantly higher. Both internodes and petioles were longer under shading in *P. anglica,* which is a hybrid of the stoloniferous *P. reptans* and the erect *P. erecta,* but internode length did not alter significantly in *P. reptans.* However, petioles of *P. reptans* grown under spectral shading were three times as long (Fig. 4b), and their mean weight twice as great as in control plants. Weight per unit length of both stolons and petioles decreased significantly when *P. anglica* and *P. reptans* were grown in shaded conditions.

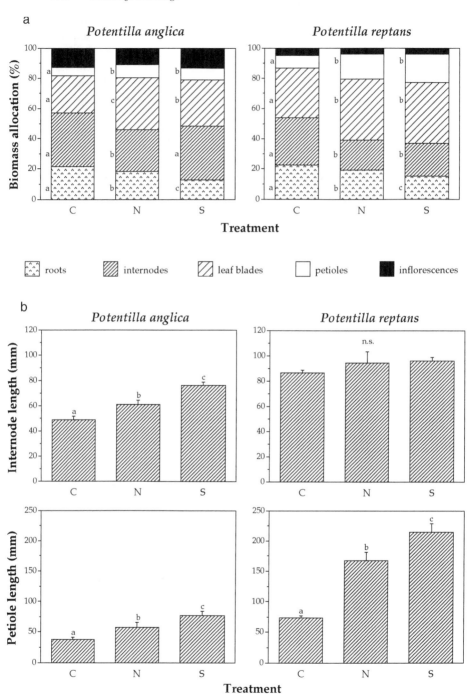

a

Potentilla anglica — *Potentilla reptans*

Biomass allocation (%)

Treatment

roots   internodes   leaf blades   petioles   inflorescences

b

Potentilla anglica — *Potentilla reptans*

Internode length (mm)

Petiole length (mm)

n.s.

Treatment

This study shows a striking elongation of vertically oriented petioles, and significant increases in the proportion of biomass allocated to these organs, in stoloniferous plants grown in shade, whereas allocation to horizontal spacing organs changed little or was reduced. Vertically oriented organs also showed a higher absolute level of plasticity than horizontal organs. In two species of *Potentilla* with erect stems, internodes were more plastic than petioles, leading Huber to speculate that analogous organs (stem internodes in the erect species and erect petioles in the stoloniferous species) respond in a similar way in poor light environments because elongation in both instances fulfils the same ecological function of enhancing light capture.

4. Response to Waterlogging   Although the R/S ratio fell with increased waterlogging in an experiment carried out by Soukupová (1994) on three clonal graminoids, all three species also showed reductions in biomass with increased flooding, making it difficult to determine whether the change in R/S ratio was due to flooding, ontogenetic state, or a combination of both.

5. Response to Severing of Rhizomes and Stolons   Clones of *Ranunculus repens* allocated more of their biomass to roots and less to stolons when connections between ramets were severed than when they remained intact (Lovett Doust, 1981b). Slade and Hutchings (1987c) compared resource allocation patterns in clonal fragments of *Glechoma hederacea* consisting of different numbers of rooted ramets. Under nutrient-rich conditions, fragments with few established ramets allocated more of their biomass to stolons, but allocation to petioles was greater in fragments with more ramets. Under nutrient-poor conditions, an increasing proportion of clone biomass was allocated to leaves as the number of ramets in the fragment increased.

Figure 4   (a) Proportional patterns of mean biomass allocation to different organs by *Potentilla anglica* and *P. reptans* when grown in full light (C), neutral shade (N), and spectral shade (S). Error bars are omitted for clarity. Different letters indicate statistically significant differences (at least $P < 0.05$) between treatments in the proportional allocation to that structure. Comparisons are made within species only. (b) Mean length ($\pm$S.E.) of internodes and petioles of *Potentilla anglica* and *P. reptans* when grown under the same treatments as in (a). Different letters above the bars indicate a significant difference between the treatments at least at $P < 0.05$. From Huber (1996).

IV. Resource Allocation Patterns in Clonal Species Growing under Heterogeneous Conditions

The responses of different parts of single clones to contrasting growing conditions have been analyzed in several experimental studies. Slade and Hutchings (1987a) compared the growth of *Glechoma hederacea* clones when the ramets of one of the two primary stolons were grown in nutrient-rich soil and those of the other stolon in nutrient-poor soil. In these "split treatments," all daughter ramets received the same conditions as the part of the clone from which they descended. Additional clones ("pure treatment" clones) were grown under uniform conditions, with both of their stolons growing in the same soil as either the low- or high-nutrient stolon of the split clones. At harvest, the two stolons of the split treatment clones differed significantly in all measured parameters, but when each stolon from the split clones was compared with the stolons receiving the matching nutrient supply in the pure clones, there were no significant differences between any morphological variables. The proportion of biomass allocated to different structural components by each half of the split treatment clones also closely matched that of the corresponding pure clones (Table I).

These results are important in the context of the present discussion because they demonstrate that different parts of the same clone adopt morphologies and resource allocation patterns which are appropriate for the local conditions. There was no evidence of compromise in growth, morphology, or biomass allocation patterns between parts of *G. hederacea* clones experiencing different growing conditions; each part became specialized for different types of environments. The lack of compromise can be attributed to the vascular architecture of *G. hederacea,* which constrains the extent to which different parts of clones can be physiologically integrated (Price *et al*, 1992), but similar observations have been made in other clonal species where such constraints appear to be absent. For example, Turkington *et al.,* (1991) have shown that different parts of *Trifolium repens* clones can produce similar biomass when grown in competition with different grasses, but that each part achieves this with a different morphology, which is presumably appropriate for the local neighborhood. This topic is discussed in greater detail elsewhere (Hutchings and Price, 1993).

Primary ramets of *Lamiastrum galeobdolon* stolons growing across habitat patches with differing PAR had shorter internodes and petioles, and smaller leaf laminae with lower SLA when in high than in low light patches (Dong, 1993). These results closely parallel those obtained for *Glechoma hederacea* (Slade and Hutchings, 1987b). Internode length in *L. galeobdolon* only responded to shade patches large enough to accommodate at least the elongating internodes on the stolon (the two most distal internodes; Dong,

1993) and the stolon apex. In contrast, internode response to high light patches was independent of patch size. High light incident on either the stolon apex or the elongating internodes resulted in shorter internodes, even when the elongating internodes had not all reached the high light patch (Dong, 1993). This may allow efficient exploitation of high light patches, an important attribute in this species of patchily shaded woodland floors.

Resource shortage caused by competition also causes changes in biomass allocation patterns, as illustrated in an experiment in which *Glechoma hederacea* clones were grown either in the absence of competition or with *Lolium perenne* (Price and Hutchings, 1996). *Lolium perenne* was either allowed to grow tall (about 8 cm at the beginning of the experiment, and about 13 cm by the end) or kept short by repeated clipping to a height of 1 cm. In the tall-grass treatment there was strong competition for both light and soil-based resources, whereas in the short-grass treatment light competition was less intense. There were significant differences between all treatments in both light transmission and R/FR ratio at the height of the stolon internodes of the clone; PAR and R/FR decreased as the height of *L. perenne* increased. Rates of production of primary ramets were the same in all treatments, enabling clones to be compared at the same stage of development, as measured by the number of plastochrons through which the clones had passed.

With short *L. perenne*, the major change in comparison with control clones was for the biomass allocated to stolons to increase by approximately 80%. Some increase in proportional allocation to petioles was also seen. This adjustment could promote lateral movement of the clone, which may enable it to grow away from the competitor, particularly as it is combined with an increase in the stolon length/weight ratio. When grown with tall *L. perenne*, the proportional allocation of biomass to petioles was increased by 133% in comparison with control clones and by 66% in comparison with clones grown with short grass competition. The clone thus increases both lateral movement and vertical growth, both of which may enable competition for light to be avoided. Proportional allocation of biomass to leaves *decreased* as competition for light intensified, from 65% in control clones to 43% in clones growing with short grass competition, and to only 27% in clones growing with tall grass competition. The adjustments in petiole height of *G. hederacea* in the tall grass competition treatment, as a result of greater proportional investment in these structures, enabled the clone to place its leaves at the top of the *L. perenne* canopy. As a consequence, there was not a significant difference between the weight achieved by clones grown with short and tall grass.

In the examples discussed so far, habitat heterogeneity was created by growing different stolons of the same clone in contrasting conditions.

However, individual stolons or rhizomes may also extend across the substrate through patches of different quality. Slade and Hutchings (1987c) analyzed the effect on stolons of *Glechoma hederacea* of moving between different quality patches. Growth was compared in homogeneous and heterogeneous treatments. Individual ramets were given either full-strength or one-eighth-strength nutrient solution, and either 100 or 25% daylight PAR. No ramets were subjected to both low nutrients and low PAR. Thus, all parts of clones in the homogeneous treatments received either high light and high nutrients (+L+N), high light and low nutrients (+L−N), or low light and high nutrients (−L+N). Primary stolons produced primary ramets at the same rate in all treatments. After production of a given number of primary ramets in all treatments, clones were either kept in the same conditions (homogeneous treatments), or all subsequently produced ramets were grown in different conditions (heterogeneous treatments). Again, all primary stolons produced ramets at the same rate. Six primary ramets were produced during the first phase of growth and a further six during the second phase. All clones therefore were at the same stage of growth in all treatments, having passed through twelve plastochrons, and having the same primary ramet age structure. Thus, direct comparisons of patterns of biomass allocation to different structural components were possible.

On the basis of dry matter production, the most favorable growing conditions were +L+N and the least favorable −L+N. The differences in dry matter production were largely caused by the extent to which secondary and tertiary stolons had grown out from primary and secondary nodes. As the primary nodes had the same age profile in all treatments, a major effect of the treatments was to change the *time* at which growth of higher order structures was initiated.

In the homogeneous treatments, a higher proportion of clone biomass was allocated to roots when nutrients were scarce (compare +L−N with +L+N), and allocation to leaves was reduced (Table IV). When light was scarce (the −L+N treatment), root allocation was low in comparison with +L+N clones. Allocation to petioles was increased, but allocation to leaves hardly differed between the +L+N and −L+N treatments (Table IV). Increasing allocation to petioles rather than to leaves may be advantageous in herbaceous species with plagiotropic stems, where the capacity to compete with neighboring herbs for light may be a valuable attribute. As reported above, petiole length increases markedly in *Trifolium repens, Glechoma hederacea,* and stoloniferous *Potentilla* species in response to competition from other herbs and in response to changes in the R/FR ratio in incident light (Solangaarachchi and Harper 1987; Thompson and Harper, 1988; Price and Hutchings, 1996; Huber, 1996). Allocation to stolons (the spacers in this clonal species) increased slightly but significantly in homogeneous

Table IV   Mean Percentage Allocation (±S.E.) of Biomass to Different Structural Components by Clones of *Glechoma hederacea* When Grown under Either Homogeneous or Heterogeneous Conditions[a]

	First part			Second part		
	+L+N	+L−N	−L+N	+L+N	+L−N	−L+N
Leaves	38 ± 1			40 ± 2		
	39 ± 1				36 ± 2	
	43 ± 1					34 ± 1*
		31 ± 1**		44 ± 2		
		32 ± 1**			31 ± 1	
			39 ± 1	42 ± 2		
			41 ± 1			39 ± 2
Petioles	13 ± 1			16 ± 1		
	12 ± 1				13 ± 1*	
	17 ± 1					26 ± 1*
		11 ± 1		13 ± 1*		
		10 ± 1***			13 ± 1***	
			30 ± 2***	17 ± 2		
			23 ± 1***			39 ± 3***
Stolons	15 ± 1			19 ± 1		
	15 ± 1				19 ± 1	
	13 ± 1					21 ± 1
		14 ± 1		18 ± 1		
		18 ± 1**			23 ± 1***	
			16 ± 1	17 ± 1		
			18 ± 1**			20 ± 2
Roots	34 ± 1			25 ± 1		
	34 ± 1				32 ± 1*	
	27 ± 2					19 ± 1*
		44 ± 2**		25 ± 2		
		40 ± 1***			33 ± 2	
			15 ± 1***	24 ± 2**		
			18 ± 1***			11 ± 2***

[a] Conditions were high light and high nutrients (+L+N), high light and low nutrients (+L−N), or low light and high nutrients (−L+N). +L, −L, +N, and −N conditions were as described in the legend to Table II. For further details, see Slade and Hutchings (1987c). Clones were grown in one set of conditions until their primary stolons had produced six ramets (First part), after which the next six primary ramets were produced under either the same (homogeneous treatments) or different (heterogeneous treatments) conditions. The rate of production of primary ramets was the same in all treatments and under all conditions. Asterisks are used to indicate significant differences between the percentage of biomass allocation to each structural component by each part of the clone, and allocation to the same component by the corresponding part of the +L+N clones (i.e., those growing continuously under favorable conditions): *, $P < 0.05$; **, $P < 0.01$; ***, $P < 0.001$. Other comparisons are described in the text.

treatments where there was resource shortage. In most cases, allocation to roots was lower in the younger part of the clones, whereas allocation to stolons and to petioles, particularly under low light conditions, was higher.

Comparisons were made between the patterns of biomass allocation in the older part of clones in homogeneous treatments (either +L−N or −L+N throughout their growth) and the older parts of clones in heterogeneous treatments that commenced growth in the corresponding conditions. There were no significant differences in proportional allocation of biomass to any structures in these older clone parts (Slade and Hutchings, 1987c), implying that basipetal movement of resources does not affect the morphology of earlier produced parts of clones, regardless of the conditions into which they grow at a later time. However, percentage allocation patterns changed in subtle ways as growing conditions altered, as can be seen by comparing, for example, clones growing from +L+N into either +L−N or −L+N conditions. There were no significant differences between the allocation patterns of the older parts of clones in either of these treatments and the older parts of clones in homogeneous +L+N treatments. However, the younger parts of clones growing from +L+N to +L−N allocated significantly more ($P < 0.05$) of their biomass to leaves and significantly less ($P < 0.01$) to stolons than the younger parts of clones in the homogeneous +L−N treatment. The younger parts of clones growing from +L+N to −L+N allocated significantly less ($P < 0.05$) of their biomass to petioles and significantly more ($P < 0.01$) to roots than the younger parts of clones in the homogeneous −L+N treatment.

These results can be interpreted as follows. Whereas basipetal translocation does not affect the morphology of older clone parts, these results imply that acropetal translocation, which is substantial in *G. hederacea* (Price *et al.*, 1992), can have a considerable impact on local morphology. Clones growing entirely under conditions of resource shortage differ from those growing in more favorable conditions by allocating more resources to structures that may help to alleviate the deficiency. For example, scarcity of light may lead to more biomass being allocated to petioles and perhaps to leaves, and scarcity of nutrients or water may lead to more allocation to roots. However, if a clone commences growth in favorable conditions and later grows into sites where there are resource shortages, the older parts of the clone support the younger parts by acropetal translocation. When this happens, the expected change in the pattern of biomass allocation is only partially expressed in the part of the clone affected by resource deficiency. In contrast, biomass allocation patterns in the younger parts of clones commencing growth in resource-poor conditions, but growing later into favorable conditions, show little difference from the patterns in the corresponding parts of clones which grow entirely in favorable conditions. This appears to be due to resource shortage for acropetal transport from the part of the clone growing under resource deficiency. The pattern of biomass allocation appropriate to the local conditions is thus expressed rapidly in response to the improved quality of the habitat. Clone parts are thus sustained by resource translocation when they enter poor conditions from good, whereas effective exploitation of good conditions following emergence from poor conditions is promoted by rapid expression of plasticity. Similar results were found by Noble and Marshall (1983) in a study reporting R/S ratios in *Carex arenaria* traversing high and low nutrient regimes.

V. Consequences of Localized Responses by Clonal Herbs to Heterogeneous Habitat Conditions

Many ecological experiments contrast growth in uniformly high or low quality habitats or in heterogeneous habitats containing good and bad patches created by varying the availability of a single resource. Alpert and Mooney (1986) have shown, using *Fragaria chiloensis*, that the ability to transport vital resources from ramets in favorable patches to those in deficient patches can change the habitat for a clonal species from a mosaic of favorable and unfavorable sites into one that is entirely favorable. Although such resource sharing benefits the recipient part of the clone, the cost to the donor ramet is often small; when connections between ramets are severed, however, those ramets in unfavorable patches are likely to die.

Unlike these simple experimental designs, real habitat patches may be deficient in any of several resources. The abundances of different resources are often negatively correlated in space (Schlesinger *et al.*, 1990; Stuefer and Hutchings, 1994). Thus, a microsite where resource A is in adequate supply may be deficient in resource B, whereas an adjacent microsite may have the opposite qualities. In such situations, no sites are wholly good, although clonal plants may make the whole habitat acceptable by establishing ramets in many microsites and exchanging acquired resources between them. Alpert and Mooney (1986) studied such a situation, providing one of two connected ramets of *Fragaria chiloensis* with a lethal scarcity of water and the other with a lethal scarcity of light. In this treatment every microsite was uninhabitable, but the clone could survive and grow well as a consequence of resource sharing.

Friedman and Alpert (1991) developed this work further in experiments on *F. chiloensis* in which connected ramets were given complementary resource deficiencies (one received high light but low nitrogen and the other low light and high nitrogen). Two experiments were carried out with nitrogen either in limiting supply or in surplus. Growth was compared with connections between ramets intact or severed. Ramets receiving high light

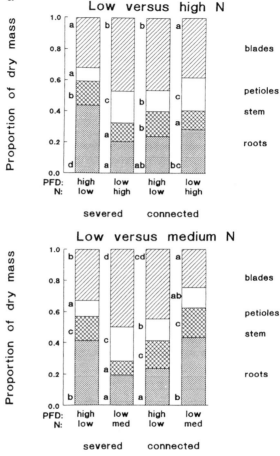

Figure 5   (a) Partitioning of dry mass between plant parts within ramets of *Fragaria chiloensis*, demonstrating the effects of vascular connections between ramets given contrasting light and soil nitrogen levels (see text). For a given plant part in a given graph, treatments with the same letter did not differ significantly in proportion of biomass allocated to a given structure (SNK or STP, *P* = 0.05). (b) Accumulation of dry mass by pairs of ramets of *Fragaria chiloensis*. Vertical lines show standard errors for total mass (upward) and for mass of each plant part (downward). For a given plant part, treatments with the same letter did not differ significantly (SNK, *P* = 0.05). From *Oecologia*, Reciprocal transport between ramets increases growth of *Fragaria chiloensis* when light and nitrogen occur in separate patches but only if patches are rich. Friedman, D., and Alpert, P., **86**, 76–80, Figs. 2 and 4, 1991, copyright Springer-Verlag.

and low nitrogen put less biomass in roots and more in leaves and petioles, and ramets receiving low light and higher nitrogen put more biomass in roots and less in leaves, when connected than when severed (Fig. 5a). Thus,

b

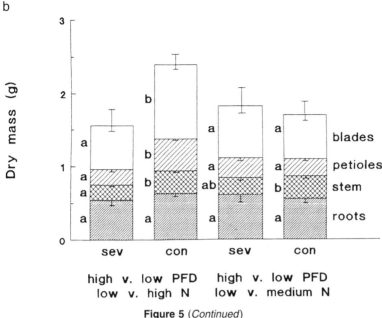

Figure 5 (*Continued*)

plant systems with single rooting sites (severed ramets) adopted a high R/S ratio when nutrients were scarce and light abundant, and a low R/S ratio when the reverse was true, as commonly observed (Brouwer, 1962; Aung, 1974; Hunt and Nicholls, 1986). In contrast, the connected ramets, which had two rooting sites, developed a high R/S ratio where nutrients were abundant and light scarce, and a low R/S ratio where the reverse was true. Each ramet thus adjusted morphologically to acquire locally abundant resources, rather than to improve the acquisition of resources that were locally scarce. This should allow the most efficient resource capture when resources can be obtained from multiple sites (Bloom *et al.*, 1985). In support of this prediction, the combined weight of the connected ramet pairs significantly exceeded that of the severed ramets, although only in the experiment with the higher nitrogen treatment (Fig. 5b).

In this experiment it appears that there is reciprocal transport of carbohydrates to the light-limited ramet and of nitrogen to the nitrogen-limited ramet. A similar situation appeared to occur when clones of *Potentilla reptans* and *P. anserina* were grown under either uniform or heterogenous light regimes in the field (Stuefer *et al.*, 1994). Clones grew larger when in heterogeneous light conditions than when in either uniform light regime, and the parts of the clone in full light and in shade both grew more than the corresponding parts in the homogeneous light and shade treatments.

It was hypothesized that the part of the clone in shade receives carbohydrates from the part in full light, and that the part of the clone in full light, which has a high evaporative demand for water, is provided with water by the shaded part. Although measurements that could support this hypothesis were not made, support comes from analysis of the R/S ratios of each part of the clones. For example, the ratio in unshaded parts was far higher when they were connected to unshaded parts than when connected to shaded parts, probably because more roots were required to satisfy the demand of a stronger connected sink for water. In addition, R/S ratios of unshaded parts were lower in heterogeneous than homogeneous treatments, suggesting that the hypothesized water shortage was being partially relieved by integration.

These examples of the benefits of clonality and of multiple sites for resource acquisition in heterogeneous habitats involve complementary bidirectional transport of resources from sites of plenty to sites of deficiency, as well as division of labor between connected ramets in sites with different resource provision qualities. Whereas ramets sited in unfavorable microsites would die or make very limited growth in isolation, connection with neighboring ramets ensures survival and possibly greater growth for the whole clone than could be made in uniform conditions. In these systems of mutually supporting modules we can begin to perceive plants as having a truly social behavior, with each module contributing in a locally appropriate manner to the good of a multimodular structure.

Transport of resources within *Glechoma hederacea* clones is predominantly acropetal rather than bidirectional, largely due to vascular architectural constraints (Stuefer and Hutchings, 1994; Price *et al.*, 1992). *Glechoma hederacea* provides perhaps the most striking example reported to date of a clonal species benefiting from habitat heterogeneity, and this is due first to very marked division of resource-acquisition duties between ramets, and second to developmental responsiveness to local habitat conditions at the level of the individual ramet. A fixed quantity of nutrients was provided in homogeneous and heterogeneous spatial arrangements within 80 by 80 cm boxes (Birch and Hutchings, 1994). Single ramets of *G. hederacea*, each producing a single stolon, were planted in each box (Fig. 6). Three treatments were used, each replicated four times:

1. A 30 cm diameter circle at the center of the box was filled with a peat-based potting compost, and the rest of the box was filled with sand mixed uniformly with the same amount of potting compost as in the central circle. Altogether, the box contained 22% compost by volume, half of which was within a central circle. The central circle covered 11% of the whole box area. This treatment is referred to as "patchy."

2. The same volumes of potting compost and sand as in the patchy treatment were mixed homogeneously throughout the box. This treatment

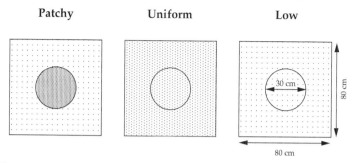

Patchy   **Uniform**   **Low**

Figure 6   Diagrammatic representation of experimental treatments in which clones of *Glechoma hederacea* were provided with nutrients in homogeneous and heterogeneous spatial configurations. Greater density of shading indicates higher concentrations of nutrients. See text and Birch and Hutchings (1994) for further explanation.

("uniform") and the patchy treatment therefore provide the same amount of nutrients in different spatial configurations. It is important to realize that the soil in the periphery of the box (nearly 90% of the total box area) was more fertile in this treatment than in the patchy treatment, although the soil in the central circle was less fertile.

3. In the final treatment ("low"), the whole box contained the same density of compost as the peripheral part of the boxes in the patchy treatment. Thus, the soil quality in this treatment and that in the patchy treatment was the same over nearly 90% of the box area.

It should be noted that the *G. hederacea* clones in the patchy and low treatments began growth in soil with the same nutrient status, whereas those in the uniform treatment began growth in more fertile soil, and this would initially have created a considerable growth advantage. This more fertile soil was available over nearly 90% of the box area in the uniform treatment. However, at harvest, the clones in the patchy treatment had produced approximately 2.5 times as much biomass as those in the uniform treatment, despite the initial growth advantage in the uniform treatment and the identical quantity of compost for growth in both treatments. Clones in the patchy treatment also produced more than 10 times the biomass of the clones in the low treatment, even though 89% of the soil volume and box area in these two treatments contained the same concentration of compost (Table V). Analysis of the proportions of biomass located in different parts of the boxes showed that clone biomass was no more concentrated in the center of the box in the patchy treatment than in the other treatments.

Two factors enabled clones in the patchy treatment to produce more biomass than clones in the other treatments. The first concerns R/S ratios.

Table V   Mean Leaf, Stolon, Root, and Total Weights (g) of Clones of *Glechoma hederacea* in Three Soil Treatments[a]

	Treatment			
	Patchy	Uniform	Low	LSD
Leaves	29.8	11.1	2.0	1.9***
Stolons	20.3	10.2	2.4	1.6***
Roots	7.5	1.6	0.9	1.1**
Total	57.6	22.9	5.3	3.4***

[a] See text for treatments. LSD indicates the least significant difference ($P < 0.05$) between treatments: **, differences between all three treatments significant ($P < 0.01$); ***, differences between all three treatments significant ($P < 0.001$).

There were no significant differences in the R/S ratios of parts of clones inside and outside the central circles in the uniform and low treatments; the R/S ratio was uniform throughout the clones under these homogeneous conditions (Table VI), although clones in the low treatment had a higher R/S ratio than clones in the uniform treatment. This parallels the change normally seen in R/S ratio in response to low nutrient supply in plants with a single rooting point. In contrast, the clones in the patchy treatment had highly significant differences in R/S ratio inside and outside the central circle (Table VI). The R/S ratio was extremely high inside the central

Table VI   Root/Shoot Ratios of Clones of *Glechoma hederacea* inside and outside the Central Circle of Three Soil Treatments[a,b]

	Treatment		
	Patchy	Uniform	Low
Inside the central circle	0.65a	0.07c	0.15b,c
Outside the central circle	0.03d	0.08c	0.19b

[a] Birch and Hutchings (1994).
[b] See text and Fig. 6 for treatments. Values in the table are medians, which were estimated by back-transformation from a two-way analysis of variance made on log-transformed measurements. Values in the table are significantly different ($P < 0.05$, using the least significant difference estimated from the analysis of variance) when they are followed by different letters.

circle of the box where nutrients were abundant, but outside the central circle, where nutrients were scarce, the ratio was extremely low, even in comparison with the equivalent parts of the low treatment clones, which were growing in soil of identical quality. As in examples quoted earlier, but to a more dramatic degree, *G. hederacea* allocated biomass within its ramets to different functions in a way that promoted acquisition of locally abundant resources.

Second, the ramets sited in the central circle in the patchy treatment established roots in the soil much earlier in their development than ramets sited in the central circle in the homogeneous treatments. Thus, they could absorb nutrients much earlier in their growth than equivalent ramets in other treatments. Moreover, whereas all ramets of clones in the homogeneous treatments, regardless of position, initiated roots at the same age, there was a considerable difference in age at which ramets rooted themselves inside and outside the central circle in the patchy treatment. Ramets inside the circle established roots approximately 5 days earlier in their development than ramets located outside the central circle. This is crucial in promoting greater growth in the heterogeneous conditions, since the amount of root produced by a ramet at a given age is strongly correlated with the age at which rooting occurs. For every day that rooting is delayed, root length at a fixed later age is reduced by 43% (Birch and Hutchings, 1994). Clearly, the ability to sense local nutrient concentration, and to use this to control the time at which rooting takes place, has a major impact on future growth, as shown by comparing the biomass accumulated by clones in the patchy and uniform treatments.

VI. Conclusions

Species that grow in the horizontal plane rarely experience homogeneous conditions. Increasingly, data are showing that clonal species develop localized patterns of biomass allocation and localized morphologies, in response to local conditions, and that such responses can be observed at a variety of structural scales. Nutrients, water, and light all provoke such plasticity. The adjustments in allocation patterns promote acquisition of heterogeneously distributed resources from the parts of the habitats where they are most abundant, and thus ramets divide the duties of acquiring each resource unequally. At the same time, changes in morphology and branching may promote exploitation of local resource abundance or escape from resource shortage. The fact that all of these responses are invoked in response to local resource supply, and that they can have significant consequences for growth, indicates that they are foraging responses (Hutchings and de Kroon, 1994).

Despite many recent advances (Caldwell and Pearcy, 1994), ecologists are still very ignorant about spatial and temporal scales of environmental heterogeneity and their consequences for plant growth and vegetation composition. Plants have clearly needed to develop ways to respond to environmental heterogeneity, and we are beginning to discover that these can be extraordinarily effective. Many questions remain unanswered, including whether different scales of heterogeneity or particular levels of contrast between patches of differing quality enable greater or lesser success in growth from a given quantity of resource. However, it appears clear that clonal species are among the best equipped to deal with the problems raised by living in a patchy world.

Acknowledgments

I gratefully acknowledge valuable help and discussions with Andy Slade, Liz Price, Colin Birch, Vince Pelling, Georgina Holdsworth, and Dushyantha Wijesinghe. Vince Pelling kindly allowed me to quote from unpublished results.

References

Alpert, P. (1991). Nitrogen sharing among ramets increases clonal growth in *Fragaria chiloensis* Ecology **72,** 69–80.
Alpert, P., and Mooney, H. A. (1986). Resource sharing among ramets in the clonal herb, *Fragaria chiloensis. Oecologia* **70,** 227–233.
Armstrong, R. A. (1982). A quantitative theory of reproductive effort in rhizomatous perennial plants. *Ecology* **63,** 982–991.
Aung, L. G. (1974). Root–shoot relationships. In "The Plant Root and Its Environment" (E. W. Carson ed.), pp. 29–61. University Press, Charlottesville, Virginia.
Austin, M. P., and Austin, B. O. (1980). Behavior of experimental plant communities along a nutrient gradient. *J. Ecol.* **68,** 891–918.
Bazzaz, F. A., and Reekie, E. G. (1985). The meaning and measurement of reproductive effort in plants. *In* "Studies in Plant Demography: A Festschrift for John L. Harper" (J. White, ed.), pp. 373–387. Academic Press, London.
Bell, A. D. (1984). Dynamic morphology: A contribution to plant population ecology. *In* "Perspectives on Plant Population Ecology" (R. Dirzo and J. Sarukhán, eds.), pp. 48–65. Sinauer, Sunderland, Massachusetts.
Bell, G., and Lechowicz, M. J. (1994). Spatial heterogeneity at small scales and how plants respond to it. *In* "Exploitation of Environmental Heterogeneity by Plants: Ecophysiological Processes Above- and Belowground" (M. M. Caldwell and R. W. Pearcy, eds.), pp. 391–414. Academic Press, San Diego.
Birch, C. P. D., and Hutchings, M. J. (1992). Analysis of ramet development in the stoloniferous herb *Glechoma hederacea* using a plastochron index. *Oikos* **63,** 387–394.
Birch, C. P. D., and Hutchings, M. J. (1994). Exploitation of patchily distributed soil resources by the clonal herb *Glechoma hederacea. J. Ecol.* **82,** 653–664.
Bloom, A. J., Chapin, F. S., and Mooney, H. A. (1985). Resource limitation in plants—An economic analogy. *Annu. Rev. Ecol. Syst.* **16,** 363–392.

Bradbury, I. K., and Hofstra, G. (1976). The partitioning of net energy resources in two populations of *Solidago canadensis* during a single developmental cycle in southern Ontario. *Can. J. Bot.* **54,** 2449–2456.

Brouwer, R. (1962). Distribution of dry matter in the plant. *Neth. J. Agric. Sci.* **10,** 361–376.

Caldwell, M. M., and Pearcy, R. W., eds. (1994). "Exploitation of Environmental Heterogeneity by Plants: Ecophysiological Processes Above- and Belowground." Academic Press, San Diego.

Clauss, M. J., and Aarssen, L. W. (1994). Patterns of reproductive effort in *Arabidopsis thaliana:* Confounding effects of size and developmental stage. *Ecoscience* **1,** 153–159.

Coleman, J. S., McConnaughay, K. D. M., and Ackerly, D. D. (1994). Interpreting phenotypic variation in plants. *Trends Ecol. Evol.* **9,** 187–191.

Cook, R. E. (1983). Clonal plant populations. *Am. Sci.* **71,** 244–253.

de Kroon, H. and Hutchings, M. J. (1995). Morphological plasticity in clonal plants: the foraging concept reconsidered. *J. Ecol.* **83,** 143–152.

de Kroon, H. and Schieving, F. (1991). Resource allocation patterns as a function of clonal morphology: A general model applied to a foraging clonal plant. *J. Ecol.* **79,** 519–530.

de Kroon, H., Stuefer, J. F., Dong, M., and During, H. (1994). On plastic and non-plastic variation in clonal plant morphology and its ecological significance. *Folia Geobotanica and Phytotaxonomica* **29,** 123–138.

d'Hertefeldt, T., and Jónsdóttir, I. S. (1994). Effects of resource availability on integration and clonal growth in *Maianthemum bifolium. Folia Geobotanica and Phytotaxonomica* **29,** 167–179.

Dong, M. (1993). Morphological plasticity of the clonal herb *Lamiastrum galeobdolon* (L.) Ehrend. and Polatschek in response to partial shading. *New Phytol.* **124,** 291–300.

Dong, M., and de Kroon, H. (1994). Plasticity in morphology and biomass allocation in *Cynodon dactylon,* a grass species forming stolons and rhizomes. *Oikos* **70,** 99–106.

Dong, M., and Pierdominici, M. G. (1995). Morphology and growth of stolons and rhizomes in three clonal grasses, as affected by different light supply. *Vegetatio* **116,** 25–32.

Eriksson, O. (1985). Reproduction and clonal growth in *Potentilla anserina* L. (Rosaceae): The relation between growth form and dry weight allocation. *Oecologia* **66,** 378–380.

Evans, G. C. (1972). "The Quantitative Analysis of Plant Growth." Blackwell, Oxford.

Friedman, D., and Alpert, P. (1991). Reciprocal transport between ramets increases growth of *Fragaria chiloensis* when light and nitrogen occur in separate patches but only if patches are rich. *Oecologia* **86,** 76–80.

Grime, J. P. (1994). The role of plasticity in exploiting environmental heterogeneity. *In* "Exploitation of Environmental Heterogeneity by Plants: Ecophysiological Processes Above- and Belowground" (M. M. Caldwell and R. W. Pearcy, eds.), pp 1–19. Academic Press, San Diego.

Harper, J. L., and Ogden, J. (1970). The reproductive strategy of higher plants. 1. The concept of strategy with special reference to *Senecio vulgaris* L. *J. Ecol.* **58,** 681–698.

Hartnett, D. C. (1990). Size-dependent allocation to sexual and vegetative reproduction in four clonal composites. *Oecologia* **84,** 254–259.

Holler, L. C., and Abrahamson, W. G. (1977). Seed and vegetative reproduction in relation to density in *Fragaria virginiana* (Rosaceae). *Am. J. Bot.* **64,** 1003–1007.

Huber, H. (1996). Plasticity of internodes and petioles in prostrate and erect *Potentilla* species. *Funct. Ecol.* **10,** 401–409.

Hunt, R., and Nicholls, A. O. (1986). Stress and the coarse control of growth and root–shoot partitioning in herbaceous plants. *Oikos* **47,** 149–158.

Hutchings, M. J., and de Kroon, H. (1994). Foraging in plants: The role of morphological plasticity in resource acquisition. *Adv. Ecol. Res.* **25,** 159–238.

Hutchings, M. J., and Price, E. A. C. (1993). Does physiological integration enable clonal herbs to integrate the effects of environmental heterogeneity? *Plant Species Biology* **8,** 95–105.

Jackson, J. B. C., Buss, L. W., and Cook, R. E. (1985). "Population Biology and Evolution of Clonal Organisms." Yale Univ. Press, New Haven, Connecticut.

Klimeš, L. and Klimešová, J. (1994). Biomass allocation in a clonal vine: effects of intraspecific competition and nutrient availability. *Folia Geobotanica & Phytotaxonomica* **29**, 237–244.

Loehle (1987). Partitioning of reproductive effort in clonal plants—A benefit-cost model. *Oikos* **49**, 199–208.

Lovett Doust, L. (1981a). Population dynamics and local specialization in a clonal perennial (*Ranunculus repens*). I. The dynamics of ramets in contrasting habitats. *J. Ecol.* **69**, 743–755.

Lovett Doust, L. (1981b). Intraclonal variation and competition in *Ranunculus repens*. *New Phytol.* **89**, 495–502.

Marshall, C. (1990). Source–sink relations of interconnected ramets. *In* "Clonal Growth in Plants: Regulation and Function" (J. van Groenendael and H. de Kroon, eds.), pp. 23–41. SPB Academic Publ., The Hague.

Mogie, M., and Hutchings, M. J. (1990). Phylogeny, ontogeny and clonal growth in vascular plants. *In* "Clonal Growth in Plants: Regulation and Function" (J. van Groenendael and H. de Kroon, eds.), pp. 3–22. SPB Academic Publ., The Hague.

Neuteboom, J. H., and Cramer, W. (1985). A comparison of the growth and dry matter distribution of couch [*Elymus repens* (L.) Gould] and perennial ryegrass (*Lolium perenne* L.) at different levels of mineral nutrition. *Neth. J. Agric. Sci.* **33**, 341–351.

Noble, J. C., and Marshall, C. (1983). The population biology of plants with clonal growth. II. The nutrient strategy and modular physiology of *Carex arenaria. J. Ecol.* **71**, 865–877.

Ogden, J. (1974). The reproductive strategy of higher plants. II. The reproductive strategy of *Tussilago farfara. J. Ecol.* **62**, 291–324.

Pavlik, B. M. (1983). Nutrient and productivity relations of the dune grasses *Ammophila arenaria* and *Elymus mollis*. II. Growth and patterns of dry matter and nitrogen allocation as influenced by nitrogen supply. *Oecologia* **57**, 233–238.

Pelling, V. (1994). Factors influencing the vegetative growth of clonal perennial species in the family Lamiaceae. D. Philos. Thesis, University of Sussex, Sussex, U.K.

Pitelka, L. F., Ashmun, J. W., and Brown, R. L. (1985). The relationship between seasonal variation in light intensity, ramet size, and sexual reproduction in natural and experimental populations of *Aster acuminatus* (Compositae). *Am. J. Bot.* **67**, 942–948.

Price, E. A. C., and Hutchings, M. J. (1996). The effects of competition on growth and form in *Glechoma hederacea. Oikos* **75**, 279–290.

Price, E. A. C., Marshall, C. and Hutchings, M. J. (1992). Studies of growth in the clonal herb *Glechoma hederacea*. I. Patterns of physiological integration. *J. Ecol.* **80**, 25–38.

Samson, D. A., and Werk, K. S. (1986). Size-dependent effects in the analysis of reproductive effort in plants. *Am. Nat.* **127**, 667–680.

Schlesinger, W. H., Reynolds, J. F., Cunningham, J. F., Huenneke, L. F., Jarrell, W. M., Virginia, R. A., and Whitford, W. G. (1990). Biological feedbacks in global desertification. *Science* **247**, 1043–1048.

Slade, A. J., and Hutchings, M. J. (1987a). The effects of nutrient availability on foraging in the clonal herb *Glechoma hederacea. J. Ecol.* **75**, 95–112.

Slade, A. J., and Hutchings, M. J. (1987b). The effects of light intensity on foraging in the clonal herb *Glechoma hederacea. J. Ecol.* **75**, 639–650.

Slade, A. J., and Hutchings, M. J. (1987c). Clonal integration and plasticity in foraging behaviour in *Glechoma hederacea. J. Ecol.* **75**, 1023–1036.

Solangaarachchi, S. M., and Harper, J. L. (1987). The effect of canopy filtered light on the growth of white clover *Trifolium repens. Oecologia* **72**, 372–376.

Soukupová, L. (1994). Allocation plasticity and modular structure in clonal graminoids in response to waterlogging. *Folia Geobotanica and Phytotaxonomica* **29**, 227–236.

Stuefer, J. F., and Hutchings, M. J. (1994). Environmental heterogeneity and clonal growth: A study of the capacity for reciprocal translocation in *Glechoma hederacea. Oecologia* **100**, 302–308.

Stuefer, J. F., During, H. J., and de Kroon, H. (1994). High benefits of clonal integration in two stoloniferous species, in response to heterogeneous light environments. *J. Ecol.* **82,** 511–518.

Thompson, K., and Stewart, A. J. A. (1981). The measurement and meaning of reproductive effort in plants. *Am. Nat.* **117,** 205–211.

Thompson, L., and Harper, J. L. (1988). The effect of grasses on the quality of transmitted radiation and its effect on the growth of white clover *Trifolium repens. Oecologia* **75,** 343–347.

Turkington, R., Sackville Hamilton, N. R., and Gliddon, C. (1991). Within-population variation in localized and integrated responses of *Trifolium repens* to biotically patchy environments. *Oecologia* **86,** 183–192.

van Andel, J., and Vera, F. (1977). Reproductive allocation in *Senecio sylvaticus* and *Chamaenerion angustifolium* in relation to mineral nutrition. *J. Ecol.* **65,** 747–758.

Waite, S., and Hutchings, M. J. (1982). Plastic energy allocation patterns in *Plantago coronopus. Oikos* **38,** 333–342.

Watson, M. A. (1984). Developmental constraints: Effect on population growth and patterns of resource allocation in a clonal plant. *Am. Nat.* **123,** 411–426.

Weiner, J. (1988). The influence of competition on plant reproduction. *In* "Plant Reproductive Ecology—Patterns and Strategies" (J. Lovett Doust and L. Lovett Doust, eds.), pp. 228–245. Oxford Univ. Press, Oxford.

8

Trade-offs between Reproduction and Growth Influence Time of Reproduction

Edward G. Reekie

I. Introduction

There is substantial intraspecific variation in time of reproduction in most monocarpic plants. For example, in the *"biennial"* *Verbascum thapsus*, reproduction may actually take place in the first, second, or third year of growth depending on latitude and successional status of the habitat (Reinartz, 1984a,b). Similarly, there is variation in the time of reproduction for a wide range of so-called biennial species (Lee and Hamrick, 1983; Silvertown, 1984; Lacey, 1986a,b; Klinkhamer *et al.*, 1991). Even in strict annuals there is often substantial seasonal variation in time of reproduction (Chiariello and Roughgarden, 1984; Geber, 1990; Thomas and Bazzaz, 1993). There are also examples of *"annual"* species that may postpone reproduction to a subsequent year (see literature cited in Kelly, 1985). In *Poa annua* for example, reproduction can be postponed to the second or even third year of growth (Law, 1979).

Although some of this variation has a genetic basis, common garden and reciprocal transplant experiments suggest that much of this variation is environmentally induced (Reinartz, 1984a,b; Lacey, 1986a, Thomas and Bazzaz, 1993). A number of different environmental factors have been shown to influence time of reproduction in plants, including nutrient availability, light quantity and quality, moisture, and temperature extremes (e.g., Reinartz 1984a,b; Lacey, 1986a; de Jong *et al.*, 1986; Klinkhamer *et al.*, 1991). To an extent, the influence of these factors on time of reproduction can be attributed to their effect on growth and, therefore, plant size (Lacey, 1986a).

Many monocarpic plants must reach a minimum critical size before reproduction can be induced (Werner, 1975; Gross, 1981; Wesselingh *et al.*, 1993, 1994). Actual induction may be triggered by vernalization or by a photoperiod signal, but many plants must reach a minimum size before these signals are effective (Klinkhamer *et al.*, 1987a,b). Presumably, this is a mechanism to ensure individuals accumulate sufficient resources to allow successful reproduction. A direct consequence of this requirement is that time of reproduction in monocarpic perennials is more closely correlated to size than age (see references cited in Lacey, 1986b). Therefore, any environmental factor that affects growth rate will also influence time of reproduction.

However, in most natural populations there is much variation in the size at which flowering occurs. For example, in *Oenothera biennis* size at reproduction can differ by several factors even among individuals within a given site in a particular year (Fig. 1). Such variation is difficult to explain if there were a single minimum size required for reproduction in a given population. Baskin and Baskin (1979) explain this variation in terms of differential growth beyond the critical size during the growing season preceding the cold period that induces flowering. An individual that reaches the minimum critical size early in the growing season and an individual that reaches the critical size at the end of the growing season would both delay flowering to the following year because of the requirement for vernalization. However, the former individual would be larger at time of flowering due to growth after reaching the minimum critical size. This can account for plants that flower at a size larger than the minimum critical size, but it cannot explain plants that remain vegetative even though they are larger than other plants that have flowered (Lacey, 1986a). Most natural populations exhibit substantial overlap in the size of flowering and nonflowering individuals (Fig. 1; see also Werner, 1975; Baskin and Baskin, 1979; Van der Meijden and Van der Waals-Kooi, 1979; Gross, 1981; Gross and Werner, 1983; Kachi and Hirose, 1985; Reinartz, 1984a,b; de Jong *et al.*, 1986), suggesting that the critical size required for reproduction does indeed vary widely. There is also direct experimental evidence that the critical size varies with environment. For example, Lacey (1986a) found that the minimum size at reproduction increased with nutrient availability in *Daucus carota*. She suggests that postponement of reproduction in a high nutrient environment would be selected for, as continued vegetative growth under fertile conditions will result in substantial increases in future reproductive output.

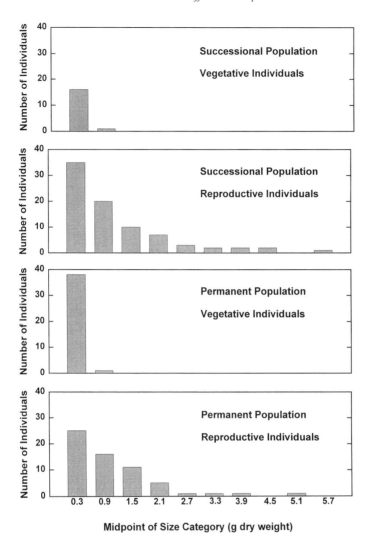

Midpoint of Size Category (g dry weight)

Figure 1 Size distribution of vegetative and reproductive individuals in two contrasting populations of *Oenothera biennis*. The successional population was on a fertile site, the second year after disturbance (plowing) and a closed canopy had developed. The permanent population was on an infertile site (gravel bed) where there was little or no shading of individual plants. Plants were harvested at time of bolting in reproductive plants. From E. G. Reekie and E. D. Parmiter (unpublished data).

II. Theoretical Models

A number of theoretical models have been developed to explain variation in time of reproduction in monocarpic plants. These models can be divided into two groups: (1) those that attempt to explain seasonal variation in annuals and (2) those that examine year to year variation in monocarpic perennials.

Early models for annuals suggest that plants should grow vegetatively for a period of time, then switch completely to reproductive growth with no overlap of the vegetative and reproductive phases (Cohen, 1971). This abrupt switch is optimal because of the compounding effect of vegetative growth; vegetative tissues contribute to future growth, and therefore reproduction should be postponed as long as possible to maximize plant size and the resources available for reproduction. Empirical data, however, show that rarely is there ever a complete switch from vegetative to reproductive growth (see literature reviewed in Rathcke and Lacey, 1985; Bazzaz *et al.*, 1987; Kozlowski, 1992). Rather, there is a gradual switch, along with a prolonged period in which vegetative and reproductive growth overlap. It has been suggested that this gradual switch can be explained by design constraints. For example, the production of new inflorescences may be morphologically linked with the production of new shoots (Watson, 1984; Geber, 1990), or it may be easier to switch on regulators of mature traits than it is to switch off juvenile traits (Poethig, 1990; Lawson and Poethig, 1995). A subsequent model (King and Roughgarden, 1982) has also shown that random variation in the length of the growing season will select for a gradual switch (see also Kozlowski, 1992).

Several models have explored the factors that may control the time of the switch from vegetative to reproductive development. Cohen (1976) demonstrated that early reproduction would be advantageous if relative growth rate decreases with plant size, if the probability of mortality increases with time, or if the reproductive structures are photosynthetic. It has also been demonstrated that high rates of vegetative tissue loss will favor an early switch to reproduction (Kozlowski, 1992). On the other hand, late reproduction will be favored if a large size is particularly advantageous. For example, large plants may be better able to compete for light (Shaffer, 1977) and for access to pollinators (Cohen, 1976). Late reproduction will also be favored if flowering uses reserves stored during the vegetative phase (Chiariello and Roughgarden, 1984; Kozlowski and Wiegert, 1986).

In models for monocarpic perennials, time of reproduction is generally explained in terms of the relationship between plant size and reproductive output, the demographic cost of postponing the time when offspring themselves will reproduce, and the risk of not surviving to a future time. These models have largely focused on why reproduction should ever be postponed beyond the first year. Hart (1977) presents a simple deterministic model that predicts that a biennial has to produce four times as many seeds as a perennial and twice as many seeds as an annual to attain the same rate of increase. This arises because biennials reproduce only once every 2 years. The fact that reproduction is often postponed beyond 2 years suggests that delaying reproduction must result in very substantial increases in reproductive output. Van der Meijden and Van der Waals-Kooi (1979) use a deterministic model to demonstrate that delayed flowering is unlikely to be selected for in growing populations but could be profitable in declining populations. This is an important point given that some monocarpic perennials occupy temporary successional habitats and disperse through time by means of seed dormancy (Reinartz, 1984a; Gross and Werner, 1982). This results in a short period of population growth followed by a long period of slow decline of the buried seed pool until the next disturbance event. Population growth rates are therefore solely dependent on the number of seeds produced and independent of the time of their production (Reinartz, 1984a). Klinkhamer and de Jong (1983) constructed a stochastic model which illustrates that, in a variable environment, delay of flowering can be profitable even in a growing population. Individuals that produce offspring that flower in different years reduce the possiblity that all of their offspring may reproduce in a "bad" year. This decreases the rate of extinction and increases population growth. De Jong *et al.*, (1989) present a stochastic model which demonstrates that the increase in reproductive output with plant size that is observed in *Cirsium vulgare* and *Cynoglossum officinale* is sufficient to explain delays in reproduction. Kachi and Hirose (1985) report similar results from a model of the population dynamics of *Oenothera glazioviana*. The model of de Jong *et al.* (1989) also demonstrates that the increase in reproductive output with plant size is more important than averaging of variable recruitment in explaining delays in reproduction.

III. Growth after the Induction of Reproduction

The objective of this chapter is to offer a possible explanation for the extensive variation in the size required for reproduction observed in many monocarpic plants. The approach taken here, however, differs from that of most previous studies in that the focus is not on growth up to the induction of flowering, but on growth after reproduction has been induced.

Given that reproduction has already been induced, it might be argued that growth after this point is irrelevant so far as time of reproduction is concerned. This is not necessarily true for two reasons. First, as discussed above, much of the growth of many monocarpic plants takes place after the induction of flowering (Rathcke and Lacey, 1985; Bazzaz *et al.*, 1987; Kozlowski, 1992). In *Oenothera biennis*, for example, the proportion of final biomass accumulated after the induction of reproduction can be as high as 50% (Reekie and Reekie, 1991). Second, there may be reliable indicators of postreproductive growth potential that could be "sensed" by an individual prior to the induction of reproduction. Nutrient availability and light quality and quantity, for example, may provide accurate indicators of future growth potential.

A number of studies have examined the effect of reproduction on growth in iterocarpic plants (e.g., Jurik, 1985; Reekie and Bazzaz, 1987a,b, 1992; Horvitz and Schemske, 1988; Lubbers and Lechowicz, 1989; Snow and Whigham, 1989; Karlsson *et al.*, 1990; Primack and Hall, 1990; Reekie, 1991; Fox and Stevens, 1991; Jackson and Dewald, 1994). Although these studies are not directly relevant to the present discussion concerning monocarps, they have shown that the impact of reproduction on growth is highly variable. Reekie and Bazzaz (1987a,b) for example, found that reproduction can either decrease or increase vegetative growth in *Agropyron repens* depending on environmental conditions. Such variation suggests that the extent to which growth continues after reproduction in monocarps is also likely to vary. Given that reproductive output will ultimately depend on the total amount of resources accumulated in growth, the extent to which growth continues after reproduction will have a major impact on the relationship between reproductive output and plant size at the time reproduction is induced. Moreover, as the above models demonstrate, the shape of this relationship is probably one of the most important factors determining time of reproduction. For example, a logical extension of these models would be that reproduction should occur at an earlier age and smaller size when reproduction has little negative effect on growth, and at a larger size and later time when reproduction has more negative effects on growth.

IV. Effect of Reproduction on Growth in Monocarpic Plants

As an example of how the impact of reproduction on growth can vary in monocarpic plants, I present some experimental data for *Oenothera biennis*. This species is a short-lived monocarpic perennial commonly found in recently disturbed or low nutrient habitats with an open canopy (Hall *et al.*, 1988). It exists as an acaulescent rosette in the vegetative state, and forms elongate stems bearing leaves, flowers, and capsules in the reproductive state. As vernalization (i.e., a cold treatment) is required, it does not normally reproduce in its first year. Although reproduction may take place in the second year, it is often postponed to the third year, or even later.

Reekie and Reekie (1991) examined the effects of reproduction on growth in *Oenothera biennis* by inducing reproduction experimentally in unvernalized vegetative rosettes through the application of gibberellic acid, a naturally occurring plant growth regulator that induces many rosette-forming plants to undergo stem elongation and to flower (Lang, 1957). Early reproduction decreases overall growth relative to vegetative controls, whereas late reproduction increases growth temporarily and has no negative effect on growth in the long term (Fig. 2). The reduced growth of plants reproducing at an early age relative to late-flowering plants is due to a decrease in leaf area ratio and the enhanced growth of older plants to a temporary increase (Fig. 3). Leaf allocation decreases with reproduction in both young and old plants, but specific leaf area in older plants increases with reproduction, temporarily overcompensating for the decrease in leaf allocation. Vegetative rosettes produce elongate leaves with a petiole, which prevents self-shading. Because of stem elongation, self-shading is less of a problem in reproductive plants. These plants produce short leaves with no petiole. Because of the low area to weight ratio of the petiole, leaves initiated after reproduction has been induced have a much higher specific leaf area than leaves produced before reproduction (Fig. 4). In older plants, the increase in specific leaf area with reproduction is sufficient to compensate for the reduction in leaf allocation associated with reproduction, and total

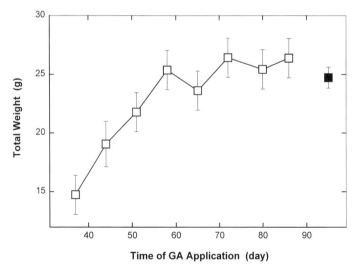

Figure 2 Effect of reproduction on growth at maturity in *Oenothera biennis*. Age at reproduction was varied by applying gibberellic acid (GA) to nonvernalized vegetative plants at various times (open symbols). Closed symbols represent vegetative controls (i.e., plants that did not receive GA). Error bars show ±1 S.E. ($n = 4$). From Reekie and Reekie (1991).

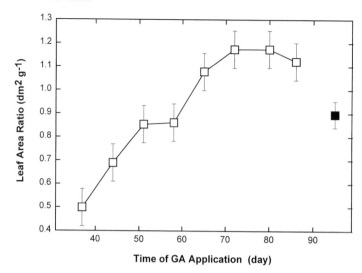

Figure 3   Effect of reproduction on leaf area ratio at flowering in *Oenothera biennis*. Age at reproduction was varied by applying gibberellic acid (GA) to nonvernalized vegetative plants at various times (open symbols). Closed symbols represent vegetative controls (i.e., plants that did not receive GA). Error bars show ± 1 S.E. ($n = 4$). From Reekie and Reekie (1991).

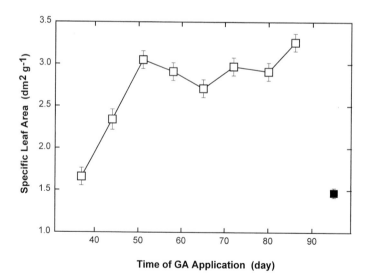

Figure 4   Effect of reproduction on specific leaf area at maturity in *Oenothera biennis*. Age at reproduction was varied by applying gibberellic acid (GA) to nonvernalized vegetative plants at various times (open symbols). Closed symbols represent vegetative controls (i.e., plants that did not receive GA). Error bars show ± 1 S.E. ($n = 4$). From Reekie and Reekie (1991).

leaf area is maintained or even increased. These effects on leaf area can compensate for the carbon allocated to reproduction, and total growth may actually increase slightly with reproduction. On the other hand, plants reproducing at an early age produce few new leaves and exhibit little change in specific leaf area (Fig. 4).

One possible explanation for these contrasting effects of reproduction on growth are differences in level of mineral reserves among plants of different ages. In monocarpic perennials, reserves of nitrogen and other mineral nutrients accumulate with plant age. For example, in *Arctium tomentosum*, 20% of the nitrogen required for reproduction is supplied by reserves accummulated in the roots during the first year of growth (Heilmeier *et al.*, 1986). Young plants therefore have fewer reserves, which are more likely to be depleted by reproduction, and require the mobilization of metabolically active nitrogen. This can deprive the photosynthetic apparatus of needed resources and so reduce carbon assimilation.

Another study (Saulnier and Reekie, 1995) tested this hypothesis by examining the effect of reproduction on photosynthetic rate, leaf area production, chlorophyll content, and nitrogen allocation in young versus old plants grown at low versus high nutrient availability. Of the various mineral nutrients, nitrogen is likely to have been the most limiting in this experiment; nutrients were supplied by watering with a complete fertilizer solution that had a relatively low nitrogen content ($15:30:15$ or $20:20:20$ N : P : K). Nitrogen availability has a marked impact on the carbon assimilation capacity of plants. Nitrogen is an integral part of the chlorophyll molecule and is part of the various structural and enzymatic proteins required for photosynthesis. As a result, leaf area production increases with the supply of nitrogen, and the rate of photosynthesis is closely correlated with leaf nitrogen content (Field and Mooney, 1986; Evans, 1989).

Reproduction was controlled experimentally by gibberellic acid applications, and measurements were made at three developmental stages: bolting, flowering, and capsule maturation. At each stage, measurements were also made on corresponding vegetative plants of the same age.

Reproduction decreases nitrogen allocation to roots and increases allocation to shoots (Fig. 5). Presumably, this reflects the fact that in many monocarpic perennials roots serve as the primary storage organ in the vegetative state and that these resources are mobilized to supply reproductive growth (Lovett Doust, 1980). The decrease in root allocation is greater at low nutrient availability (Fig. 5), suggesting that the depletion of reserves is more rapid when N is less available. Reproduction increases leaf area, and, at bolting, the magnitude of this increase is greater for older plants and for plants grown at high nutrient availability (Fig. 6). This means that older plants and plants grown at high nutrient availability are better able to continue growth during reproduction. Reproduction decreases chlorophyll

200   *Edward G. Reekie*

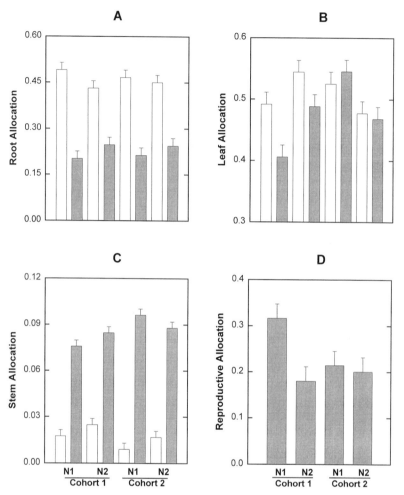

Figure 5 Proportion of total plant nitrogen allocated to (A) roots, (B) leaves, (C) stems, and (D) reproductive parts in vegetative (open bars) versus reproductive (shaded bars) plants at capsule maturation. There were two cohorts, grown at either low (N1) or high (N2) nutrient availability. Error bars depict one standard error of a single treatment mean. From Saulnier and Reekie (1995).

and nitrogen content of the leaves, but the decreases are less or nonexistent for plants grown at high nutrient availability. As a result, photosynthetic rate does not necessarily decrease with reproduction in older plants grown at high nutrient availability, and in the latter part of the experiment there are actually slight increases (Fig. 7).

On the basis of the above studies, it appears that differences among *Oenothera biennis* individuals in the effect of reproduction on carbon gain

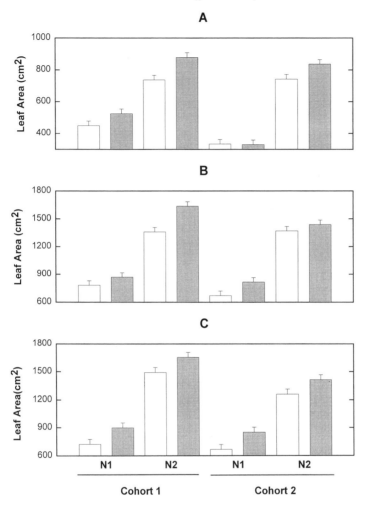

Figure 6 Leaf area of vegetative (open bars) versus reproductive (shaded bars) plants at three reproductive stages: (A) bolting, (B) flowering, and (C) capsule maturation. There were two cohorts, grown at either low (N1) or high (N2) nutrient availability. Error bars depict one standard error of a single treatment mean. From Saulnier and Reekie (1995).

are related to differences in extent of nutrient reserves. Older plants and plants grown at high nutrient availability have greater nutrient reserves on which to draw when reproduction is initiated. Reproduction in younger plants grown at lower nutrient availability will rapidly deplete nutrient reserves, and nutrients that are part of the photosynthetic apparatus (e.g., the nitrogen within the chlorophyll molecule and Rubisco) will have to be mobilized to supply reproductive structures. Reproduction in this latter

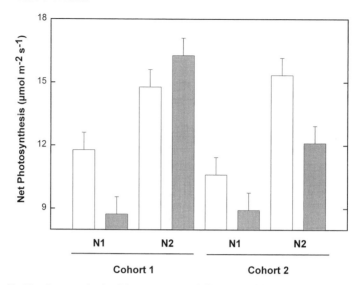

Figure 7   Net photosynthesis of the most recent fully emerged leaf of vegetative (open bars) versus reproductive (shaded bars) plants at capsule maturation. There were two cohorts, grown at either low (N1) or high (N2) nutrient availability. Error bars depict one standard error of a single treatment mean. From Saulnier and Reekie (1995).

case will therefore have more of a detrimental effect on photosynthetic rate and leaf area production. Studies with several other species support this conclusion. Heilmeier and co-workers examined lifetime carbon and nitrogen balance in the monocarpic perennial *Arctium tomentosum* (Heilmeier *et al.*, 1986; Heilmeier and Whale, 1987). They suggest that the decrease in vegetative biomass they observed after seed set is a direct consequence of nitrogen reallocation and the effect this has on photosynthesis. Sinclair and de Wit (1975) examine the nitrogen requirements for seed production in several annual crop species and relate this to the rate at which nitrogen can be taken up from the soil. They found that the nitrogen requirement for seed production usually exceeds the uptake capacity of the plant, and nitrogen has to be scavenged from vegetative structures to support reproductive growth. This depletion of vegetative nitrogen results in the "self-destruction" of the plant when nitrogen availability is low.

The situation described above is not unique. There are other environmental conditions that are likely to affect the trade-off between reproduction and growth. Studies with iterocarpic plants have shown that the trade-off between reproduction and growth can vary widely among environments. Jurik (1985) used a model to integrate data from field studies of phenology, biomass, demography, and environment with laboratory studies of CO_2 exchange to estimate cost of propagules in *Fragaria*. On the basis of total investment in reproductive structures and respiration minus photosynthetic gain of the reproductive structures, it was determined that cost per propagule is lowest in open habitats, and higher and more variable in closed habitats. Reekie and Bazzaz (1987a,b) used photoperiod manipulations to control reproduction in *Agropyron repens* and compared vegetative and reproductive individuals of the same genotype to estimate the cost of reproduction. It was found that reproduction has very different effects on growth depending on the level of light and nitrogen availability. Plants grown at high levels of these resources compensate for the carbon allocated to reproduction through direct photosynthesis of the reproductive structures and enhanced leaf photosynthesis. Lubbers and Lechowicz (1989) examined the relationship between reproduction and storage reserves in *Trillium grandiflorum* and found that the negative effects of reproduction increase with level of defoliation. Similarly, Primack and Hall (1990) found that reproduction in *Cypripedium acaule* has more negative effects on growth when combined with defoliation. They also found that as the frequency of reproduction increases, the negative impact of reproduction in any one year increases.

The above studies suggest that, in addition to variation in nutrient supply, variation in frequency of reproduction, defoliation, and light availability all influence the trade-off between reproduction and growth. In the specific case of *O. biennis*, light availability is probably particularly important. Because vegetative plants form flat rosettes and reproduction is associated with stem elongation, reproduction will likely have different consequences for growth in sites with open versus closed canopies. Presumably, the stem elongation associated with reproduction would increase the ability of plants to compete for light in a closed canopy and, therefore, would tend to increase growth. This was the question addressed in a garden experiment in which unvernalized vegetative rosettes were transplanted into weeded versus unweeded plots (E. G. Reekie and J. Y. C. Reekie, unpublished data). Half the plants in each set of plots were then treated with gibberellic acid to induce flowering, and growth of vegetative versus reproductive individuals was compared after a 4-week period. Reproduction has no significant effect on growth in the weeded plots and increases growth in the unweeded plots (Table I).

V. Implications for Allocation Patterns

The fact that reproduction has different effects on growth depending on the level of nitrogen and light availability has implications for under-

Table I   Total Biomass (g) of Vegetative and
Reproductive Individuals of *Oenothera biennis* in
Weeded Garden Plots versus a Second-Year
Successional Field[a]

	Vegetative	Reproductive
Weeded plots	20.53 ± 2.09	21.99 ± 2.01
Successional plots	2.04 ± 0.24	3.02 ± 0.36

[a] Sites were immediately adjacent to each other and on the same soil type. All individuals were from a common seed source and were germinated in a glasshouse and transplanted to the respective sites as unvernalized rosettes. Half the plants at each site were induced to reproduce by the application of gibberellic acid. Plants were harvested 4 weeks after bolting occurred in the reproductive plants. From E. G. Reekie and J. Y. C. Reekie (unpublished data).

standing the factors controlling time of reproduction. For example, because reproduction in *Oenothera biennis* has negative effects on carbon gain at low nutrient availability, a larger minimum critical size will be required to produce a given level of reproductive output, than at high nutrient availability. Therefore, in low nutrient environments, reproduction may be postponed to allow plant size to increase to a larger size than would be the case in high nutrient environments. Similarly, if competition for light is important, plants may reproduce at a smaller size because reproduction may actually enhance vegetative growth through stem elongation.

To test these predictions, we examined the size at which reproduction occurs in two contrasting populations of *Oenothera biennis* (E. G. Reekie and E. D. Parmiter, unpublished data). *Oenothera biennis* can be found in two very different habitats: (1) sites with low fertility such as gravel pits, where complete canopy closure never occurs due to nutrient shortage and *O. biennis* forms more or less permanent populations, and (2) more fertile successional fields in which *O. biennis* colonizes in the first year after disturbance, completes its life cycle, and then is eliminated from the site as canopy closure occurs (Gross and Werner, 1982). *Oenothera biennis* requires a high red/far red light ratio for seed germination, and this effectively limits recruitment to sites where bare ground is present (Gross, 1985). Given that nutrient availability is greater and light availability is lower in successional sites than in the permanent populations, we would predict that the trade-off between reproduction and vegetative growth would be lower in the successional sites. As a result, the minimum critical size required for reproduction should be lower in successional sites. A harvest of vegetative and reproductive individuals from both types of populations support this prediction (Fig. 1). Although, on average, plants are larger at reproduction in the successional site, there is also greater variation in size at reproduction at this site, and in the smallest size category (0 to 0.6 g) a larger proportion of plants reproduces in the successional site (69%) than in the permanent site (40%; $p < 0.002$ based on a 2×2 contingency table).

If individuals modify the size at which they reproduce depending on environment as suggested by the above study, then what are the physiological mechanisms involved? There are known physiological mechanisms that enable plants to sense variation in both light and nutrient environment and that would allow them to adjust time of reproduction in an appropriate fashion. The role of phytochrome in sensing changes in the red/far red ratio brought about by canopy closure and its subsequent effects on germination, stem elongation, branching patterns, and leaf morphology have been well studied (see literature cited in Morgan and Smith, 1981). The role of phytochrome in measuring day length and therefore controlling time of reproduction in day length-sensitive species has also been well studied (Morgan and Smith, 1981). The above study suggests that, in some species, it may also play a role in controlling time of reproduction through changes in light quality associated with canopy closure. The effect of nutrient availability on time of flowering, on the other hand, is probably mediated through direct hormonal changes. One indication of this interaction is provided by studies of genetic mutants. It has been shown that mutants for gibberellic acid production in *Arabidopsis thaliana* differ in their flowering response when grown in different nutrient environments (van Tienderen *et al.*, 1996).

The concept that variation in the trade-off between reproduction and growth can explain patterns of reproductive allocation has not received the attention it deserves by plant biologists. Attempts to explain variation in reproductive allocation patterns have largely focused on the demographic characteristics of the populations involved (i.e., variation in survivorship schedules and how this may select for different life histories). Such studies clearly illustrate the importance of survivorship patterns in explaining genetic differentiation in reproductive timing as well as other aspects of reproductive allocation (e.g., Law, 1979). However, most of the life history variation observed in the field has an environmental, rather than a genetic basis (see literature cited above as well as that cited in Willson, 1983; Lotz and Blom, 1986; Sultan, 1987; Bazzaz and Ackerly, 1992). Traditional life history theory was developed to explain genetic differentiation among populations and does not directly address environmental variation (Reznick, 1985). It can only be used to explain phenotypic plasticity in life history characteristics if an individual can "predict" the probability of its continued survival, as well as the likely success of its offspring in different environments (Willson, 1983). Although this is possible (i.e., survivorship may be closely correlated with environmental cues such as resource availability), it is more parsimonious to explain phenotypic variation in terms of direct environmental effects. The *Oenothera biennis* study described in this chapter provides one example of such direct effects. Simple physiological mechanisms such as those that explain the varying trade-off between reproduction and growth in *Oenothera biennis* could explain much of the environmental variation found in reproductive allocation patterns in plants.

VI. Conclusions

1. There is a great deal of intraspecific variation in reproductive timing in monocarpic plants.

2. Most of this variation has an environmental, rather than a genetic basis.

3. Some of this variation can be explained by the effect of the environment on growth, as many monocarps must reach a minimum critical size before they can reproduce.

4. However, the minimum size required for reproduction also displays a great deal of intraspecific variation.

5. Theoretical models suggest that the relationship between plant size at reproductive induction and reproductive output is crucial in determining time of reproduction.

6. The relationship between size at reproductive induction and reproductive output will be strongly influenced by the extent to which vegetative growth continues after reproduction is initiated.

7. The effect of reproduction on vegetative growth varies substantially depending on environment. In *Oenothera biennis,* for example, reproduction has little negative effect on vegetative growth when plants are in an environment with high nutrient availability and are competing for light.

8. Variation in the trade-off between reproduction and growth may explain much of the phenotypic variation in reproductive timing in monocarpic plants, with plants reproducing earlier when reproduction has little negative effect on growth and reproducing later when reproduction has more marked detrimental effects.

References

Bazzaz, F. A., and Ackerly, D. D. (1992). Reproductive allocation and reproductive effort in plants. *In* "Seeds: The Ecology of Regeneration in Plant Communities" (M. Fenner, ed.), pp. 1–26. C.A.B. International, Wallingford, Oxon, U.K.

Bazzaz, F. A., Chiariello, N. R., Coley, P. D., and Pitelka, L. F. (1987). Allocating resources to reproduction and defense. *BioScience* **37,** 58–67.

Baskin, J. M., and Baskin, C. M. (1979). Studies on the autecology and population biology of the weedy monocarpic perennial, *Pastinaca sativa. J. Ecol.* **67,** 601–610.

Chiariello, N., and Roughgarden, J. (1984). Storage allocation in seasonal races of an annual plant: Optimal versus actual allocation. *Ecology* **65**, 1290–1301.

Cohen, D. (1971). Maximizing final yield when growth is limited by time or by limiting resources. *J. Theor. Biol.* **33**, 299–307.

Cohen, D. (1976). The optimal timing of reproduction. *Am. Nat.* **110**, 801–807.

de Jong, T. J., Klinkhamer, P. G. L., and Prins, A. H. (1986). Flowering behavior of the monocarpic perennial *Cynoglossum officinale* L. *New Phytol.* **103**, 219–229.

de Jong, T. J., Klinkhamer, P. G. L., Geritz, S. A. H., and Van der Meijden, E. (1989). Why biennials delay flowering: An optimization model and field data on *Cirsium vulgare* and *Cynoglossum officinale. Acta Bot. Neerl.* **38**, 41–55.

Evans, J. R. (1989). Photosynthesis and nitrogen relationships in leaves of C₃ plants. *Oecologia* **78**, 9–19.

Field, C., and Mooney, H. A. (1986). The photosynthesis–nitrogen relationship in wild plants. *In* "On the Economy of Form and Function" (T. J. Givinsh, ed.), pp.25–55. Cambridge Univ. Press, Cambridge.

Fox, J. F., and Stevens, G. C. (1991). Costs of reproduction in a willow: Experimental responses vs. natural variation. *Ecology* **72**, 1013–1023.

Geber, M. A. (1990). The cost of meristem limitation in *Polygonum arenastrum:* Negative genetic correlations between fecundity and growth. *Evolution* **44**, 799–819.

Gross, K. L. (1981). Predictions of fate from rosette size in 4 biennial plant species *Verbascum thapsus, Oenothera biennis, Daucus carota* and *Tragopogon dubius. Oecologia* **20**, 197–201.

Gross, K. L. (1985). Effects of irradiance and spectral quality on the germination of *Verbascum thapsus* L. and *Oenothera biennis* L. seeds. *New Phytol.* **101**, 531–541.

Gross, K. L., and Werner, P. A. (1982). Colonizing abilities of "biennial" plant species in relation to ground cover: Implications for their distributions in a successional sere. *Ecology* **63**, 921–931.

Gross, K. L., and Werner, P. A. (1983). Probabilities of survival and reproduction relative to rosette size in the common burdock (*Arctium minus:* Compositae). *Am. Midl. Nat.* **109**, 185–193.

Hall, I. V., Steiner, E., Threadgill, P., and Jones, R. W. (1988). The Biology of Canadian weeds. 84. *Oenothera biennis* L. *Can. J. Plant Sci.* **68**, 163–173.

Hart, R. (1977). Why are biennials so few? *Am. Nat.* **111**, 792–799.

Heilmeier, H., and Whale, D. M. (1987). Carbon dioxide assimilation in the flowerhead of *Arctium. Oecologia* **73**, 109–115.

Heilmeier, H., Schulze, E. D., and Whale, D. M. (1986). Carbon and nitrogen partitioning in the biennial monocarp *Arctium tomentosum* Mill. *Oecologia* **70**, 466–474.

Horvitz, C. C., and Schemske, D. W. (1988). Demographic cost of reproduction in a neotropical herb: An experimental field study. *Ecology* **69**, 1741–1745.

Jackson, L. L., and Dewald, C. L. (1994). Predicting evolutionary consequences of greater reproductive effort in *Tripsacum dactyloides,* a perennial grass. *Ecology* **75**, 627–641.

Jurik, T. W. (1985). Differential costs of sexual and vegetative reproduction in wild strawberry populations. *Oecologia* **66**, 394–403.

Kachi, N., and Hirose, T. (1985). Population dynamics of *Oenothera glazioviana* in a sand-dune system with special reference to the adaptive significance of size-depenent reproduction. *J. Ecol.* **73**, 887–901.

Karlsson, P. S., Svensson, B. M., Carlsson, B. A., and Nordell, K. O. (1990). Resource investment in reproduction and its consequences in three *Pinguicula* species. *Oikos* **59**, 393–398.

Kelly, D. (1985). On strict and facultative biennials. *Oecologia* **67**, 292–294.

Klinkhamer, P. G. L., and de Jong, T. J. (1983). Is it profitable for biennials to live longer than two years? *Ecol. Modell.* **20**, 223–232.

Klinkhamer, P. G. L., de Jong, T. J., and Meelis, E. (1987a). Delay of flowering in the 'biennial' *Cirsium vulgare:* Size effects and devernalization. *Oikos* **49**, 303–308.

Klinkhamer, P. G. L., de Jong, T. J., and Meelis, E. (1987b). Life-history variation and the control of flowering in short-lived monocarps. *Oikos* **49**, 309–314.

Klinkhamer, P. G. L., de Jong, T. J., and Meelis, E. (1991). The control of flowering in the monocarpic perennial *Carlinia vulgaris*. *Oikos* **61**, 88–95.

King, D., and Roughgarden, J. (1982). Graded allocation between vegetative and reproductive growth for annual plants in growing season of random length. *Theoretical Population Biology* **22**, 1–16.

Kozlowski, J. (1992). Optimal allocation of resources to growth and reproduction: Implications for age and size at maturity. *Trends Ecol. Evol.* **7**, 15–19.

Kozlowski, J., and Wiegert, R. G. (1986). *Theoretical Population Biology* **29**, 16–37.

Lacey, E. P. (1986a). The genetic and environmental control of reproductive timing in a short-lived monocarpic species *Daucus carota* (Umbelliferae). *J. Ecol.* **74**, 73–86.

Lacey, E. P. (1986b). Onset of reproduction in plants: Size versus age-dependency. *Trends Ecol. Evol.* **1**, 72–75.

Lang, A. (1957). The effect of gibberellin on flower formation. *Proc. Natl. Acad. Sci. U.S.A.* **43**, 709–711.

Law, R. (1979). The costs of reproduction in an annual meadow grass. *Am. Nat.* **113**, 3–16.

Lawson, E. J. R., and Poethig, R. S. (1995). Shoot development in plants: Time for a change. *Trends Genet.* **11**, 263–268.

Lee, J. M., and Hamrick, J. L. (1983). Demography of two natural populations of musk thistle (*Carduus nutans*). *J. Ecol.* **71**, 923–936.

Lotz, L. A. P., and Blom, C. W. P. M. (1986). Plasticity in life history traits of *Plantago major* L. ssp. *pleiosperma* Pilger. *Oecologia* **69**, 25–30.

Lovett Doust, J. (1980). Experimental manipulation of patterns of resource allocation in the growth cycle of *Smyrnium olusatrum*. *Biol. J. Linn. Soc.* **13**, 155–166.

Lubbers, A. E., and Lechowicz, M. J. (1989). Effects of leaf removal on reproduction vs. belowground storage in *Trillium grandiflorum*. *Ecology* **70**, 85–96.

Morgan, D. C., and Smith, H. (1981). Non-photosynthetic responses to light quality. *In* "Encyclopedia of Plant Physiology" (O. L. Lange, P. S. Nobel, C. B. Osmond, and H. Ziegler, Eds.). Vol. 12A, pp. 109–134. Springer-Verlag, Berlin.

Poethig, R. S. (1990). Phase change and the regulation of shoot morphogenesis in plants. *Science* **250**, 923–930.

Primack, R. B., and Hall, P. (1990). Costs of reproduction in the pink lady's slipper orchid: A four-year experimental study. *Am. Nat.* **136**, 638–656.

Rathcke, B., and Lacey, E. P. (1985). Phenological patterns of terrestrial plants. *Annu. Rev. Ecol. Syst.* **16**, 179–214.

Reekie, E. G. (1991). Cost of seed versus rhizome production in *Agropyron repens*. *Can. J. Bot.* **69**, 2678–2683.

Reekie, E. G., and Bazzaz, F. A. (1987a). Reproductive effort in plants. 1. Carbon allocation to reproduction. *Am. Nat.* **129**, 876–896.

Reekie, E. G., and Bazzaz, F. A. (1987b). Reproductive effort in plants. 3. Effect of reproduction on vegetative activity. *Am. Nat.* **129**, 907–919.

Reekie, E. G., and Bazzaz, F. A. (1992). Cost of reproduction in genotypes of two congeneric plant species with contrasting life histories. *Oecologia* **90**, 21–26.

Reekie, E. G., and Reekie, J. Y. C. (1991). The effect of reproduction on canopy structure, allocation and growth in *Oenothera biennis*. *J. Ecol.* **79**, 1061–1071.

Reinartz, J. A. (1984a). Life history variation of common mullein (*Verbascum thapsus*) I. Latitudinal differences in population dynamics and timing of reproduction. *J. Ecol.* **72**, 897–912.

Reinartz, J. A. (1984b). Life history variation of common mullein (*Verbascum thapsus*) III. Differences among sequential cohorts. *J. Ecol.* **72**, 927–936.

Reznick, D. (1985). Costs of reproduction, an evaluation of the empirical evidence. *Oikos* **44**, 257–267.

Saulnier, T. P., and Reekie, E. G. (1995). Effects of reproduction on nitrogen allocation and carbon gain in *Oenothera biennis. J. Ecol.* **83**, 23–29.

Schaffer, W. M. (1977). Some observations on the evolution of reproductive rate and competitive ability in flowering plants. *Theoretical Population Biology* **11**, 90–104.

Silvertown, J. (1984). Death of the elusive biennial. *Nature (London)* **310**, 271.

Sinclair, T. R., and de Wit, C. T. (1975). Photosynthate and nitrogen requirements for seed production by various crops. *Science* **189**, 565–567.

Snow, A. A., and Whigham, D. F. (1989). Costs of flower and fruit production in *Tipularia disclor* (Orchidaceae). *Ecology* **70**, 1286–1293.

Sultan, S. (1987). Evolutionary implications of phenotypic plasticity in plants. *Evol. Biol.* **21**, 127–178.

Thomas, S. C., and Bazzaz, F. A. (1993). The genetic component in plant size hierarchies: Norms of reaction to density in a *Polygonum* species. *Ecol. Monogr.* **63**, 231–249.

Van der Meijden, E., and Van der Waals-Kooi, R. E. (1979). The population ecology of *Senecio jacobaea* in a sand dune system. I. Reproductive strategy and the biennial habit. *J. Ecol.* **67**, 131–153.

Van Tienderen, P. H., Hammad, I., and Zwaal, F. C. (1996). Pleiotropic effects of flowering time genes in the annual crucifer *Arabidopsis thaliana* (Brassicaceae). *Am. J. Bot.* **83**, 169–174.

Watson, M. A. (1984). Developmental constraints: Effect on population growth and patterns of resource allocation in a clonal plant. *Am. Nat.* **123**, 411–426.

Werner, P. A. (1975). Predictions of fate from rosette size in teasel (*Dipsacus fullonum* L.). *Oecologia* **20**, 197–201.

Wesselingh, R. A., de Jong, T. J., Klinkhamer, P. G. L., Van Dijik, M. J., and Schlatmann, E. G. M. (1993). Geographical variation in threshold size for flowering in *Cynoglossum officinale. Acta Bot. Neerl.* **42**, 81–91.

Wesselingh, R. A., Klinkhamer, P. G. L., de Jong, T. J., and Schlatmann, E. G. M. (1994). A latitudinal cline in vernalization requirement in *Cirsium vulgare. Ecography* **17**, 272–277.

Willson, M. F. (1983). "Plant Reproductive Ecology." Wiley, New York.

9

Size-Dependent Allocation to Male and Female Reproduction

Peter G. L. Klinkhamer and T. J. de Jong

I. Introduction

Life-history theory is a still growing field of interest to evolutionary biologists. Basically life-history theory is concerned with three questions (Stearns, 1992; Roff, 1992):

1. At what age or size does reproduction start?
2. What proportion of resources or biomass is allocated to growth, storage, and reproduction?
3. How are resources allocated to reproduction divided over male and female function?

All these questions are in fact allocation problems: what are the options for dividing time and resources over different functions, what are the consequences for growth and reproduction, and how do these translate into fitness? Life-history theory not only seeks to describe the different patterns but also attempts to find an evolutionary explanation for them. In finding such explanations knowledge about the physiological mechanisms behind variation in allocation patterns plays an important role, and this closely links physiological ecology and life-history theory.

In this chapter we address the third question and restrict ourselves to hermaphrodite monocarpic plants. In contrast to the animal kingdom where individuals of most species have separate sexes, in most plant species individuals combine both sexes usually even within the same flower, and

72% of all plant species are hermaphrodite (Yampolsky and Yampolsky, 1922). With respect to sex allocation, hermaphrodites have for a long time been considered to be invariant in their behavior because all their flowers have functional pollen and ovules. Most attention of evolutionary biologists has focused on the exceptions, for example, the evolution of dioecy (e.g., Thomson and Brunet, 1990) or the maintenance of male steriles within populations of gynodioecious species (e.g., Saumitou-Laprade *et al.*, 1994), rather than on the rule: hermaphroditism.

Horovitz (1978), Loyd and Bawa (1984), Charnov (1984), Goldman and Willson (1986), and Brunet (1992) recognized that hermaphrodite individuals are not necessarily equally effective male and female parents and that this should have consequences for the evolution of allocation patterns to male and female reproduction within this group of plants. Some hermaphrodite individuals do not set seed at all. Until the late seventies, such individuals were considered to be failures that had, for instance, insufficient levels of pollination, and many times such plants were not even taken into account when studying the reproductive biology of plants. In *The Reproductive Capacity of Plants*, Salisbury (1942) wrote, "we can for the presence ignore the condition where the individuals are so starved and depauperate that not only capsule production is reduced to one or two but the lack of nutrition leads to the abortion of seeds . . . since obviously if they represented any appreciable proportion the species would not survive." Although the latter is true we might also say from an evolutionary point of view that these plants shifted their allocation from female to male reproduction and in fact maximized their fitness by aborting seeds and trying to be a successful father.

Hermaphrodite plants may change their sex allocation in two ways: (1) by varying the number of pollen grains and ovules per flower and (2) by varying the ratio at which seeds and flowers (with pollen) are produced. Most hermaphrodite species produce many more flowers than would be necessary for the production of seeds (Stephenson, 1981; Sutherland and Delph, 1984; Sutherland, 1986a,b; Ayre and Whelan, 1989), and it has been suggested that perfect flowers are produced that function solely as pollen donors.

Usually it is assumed in sex allocation theory that resource limitation causes a trade-off between allocation to male and female function. Selection should favor a shift toward the reproductive function that produces the greatest fitness per unit of investment (Charnov and Bull, 1977; Charlesworth and Charlesworth, 1981; Charnov, 1982; Lloyd, 1984, 1988; Lloyd and Bawa, 1984; Goldman and Willson, 1986; Iwasa, 1990; Stanton and Galloway, 1990; Spalik, 1991; Morgan, 1993).

To answer the question if a particular allocation pattern to male (flowers with pollen) and female reproduction (seeds) is adaptive it is essential to know the so-called fitness gain curves, i.e., the curves that relate the investment in male or female function to the fitness gained by that function, that is, the number of offspring produced as male or female that eventually will reproduce themselves. By definition fitness includes not only seed production or the number of seeds fathered but also the success of those seeds. The success as a male or female parent will depend on genotype, environmental conditions (soil nutrient or water status, pollination level, etc.), the timing of flowering, and, of course, the investment made (e.g., Charnov and Bull, 1985; Charnov, 1984). The question is if plants can adjust allocation patterns to their expected success. Several studies reported that plants change their gender in relation to environmental conditions. Lloyd and Bawa (1984) showed how selection may lead to adjusment of allocation patterns in different environments. Variation in sex allocation with time of the season was found by Ashman and Baker (1992). De Jong and Klinkhamer (1989a) discussed the relationship between gender and plant size. In this chapter we extend our results.

II. Fitness Gain Curves in Animal-Pollinated Plants

We discuss simultaneously hermaphrodite, monocarpic species. By definition monocarpic species reproduce only once and invest all available resources in reproduction. Analogous to fitness gain curves, plant mass can then be used as a measure closely related to the total reproductive investment. In contrast to most animal species, the size or mass of reproducing plants usually varies strongly within populations; a difference by a factor 100 or more is not uncommon. For animal-pollinated plants, plant size will affect male and female fitness in several ways. We only discuss here the processes that we consider most important. Another, yet incomplete, overview is published in de Jong and Klinkhamer (1994).

A. Female Performance (Seed Production)

Large plants produce more flowers and attract more pollinators that in turn may lead to more or better quality seeds, which potentially would make large plants "good" mothers relative to their size. For a large number of species, however, it has been shown that seed production is not limited by pollination (Willson and Burley, 1983; but see Burd, 1994). Also, the success of a seed is not likely to be equal for that produced on a large or a small plant. Dudash (1991) found for *Sabatia angularis* that larger plants produced 20–40 times the number of seeds compared to small plants, although the seed dispersal distributions were not significantly different. Consequently the same amount of space and resources is divided over a greater number of offspring in large plants. Although this has not been empirically established, we would expect on the basis of simple allometric considerations that, in general, seedling density is higher around larger plants and that seedlings from large plants suffer more from local resource competition. If this is the case, the female fitness curve levels off with plant mass (Fig. 1).

B. Male Performance (Number of Seeds Sired)

Again, flowers on larger plants may receive more pollinator visits, and more pollen may be removed from the anthers. However, pollinators will visit more flowers in succession on the same plant (Geber, 1985; Klinkhamer *et al.*, 1989; Klinkhamer and de Jong, 1990), and the pollen removed from the anthers is deposited on flowers of the same plant and thus less efficiently exported to other plants (de Jong *et al.*, 1992; Klinkhamer *et al.*, 1994a). Large plants will have higher levels of geitonogamy, i.e., the pollination of flowers by neighboring flowers on the same plant (Dudash, 1991; de Jong *et al.*, 1992). It is easy to see that this will reduce male fitness if the pollen deposited within the plant is wasted (self-incompatible plants), if self pollen is less competitive than outcross pollen (Aizen *et al.*, 1990), or when selfed seeds suffer from inbreeding depression (self-compatible plants). However, even with complete self-compatibility and without any inbreeding depression, the male fitness curve levels off if the percentage selfing increases with plant size (de Jong *et al.*, 1997). The level of geitonogamy can be considerable (Pleasants, 1991; Robertson, 1992; de Jong *et al.*, 1993). The male fitness curve is therefore also expected to level off with plant mass, or number of flowers produced, at least in situations where pollinators are not scarce.

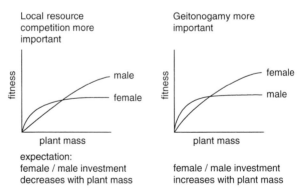

Figure 1   Hypothetical curves describing the relationship between plant size and male and female fitness under conditions of strong local resource competition and strong local mate competition.

If the male fitness curve levels off more quickly with plant size than the female fitness curve, large plants are more successful as a mother than as a father (Fig. 1). If plants can change their sex allocation pattern it would be adaptive to shift their allocation pattern to the function that gives the highest fitness return per investment. This would be seed production in large plants and flower (with pollen) production in small plants (Charnov, 1982). If, on the other hand, the female fitness curve levels off more quickly than the male curve we would expect the relative investment in seeds to decrease with plant size (Fig. 1).

So far, there are no empirical data on fitness gain curves to support either of the two possibilities. It is one of the challenges of people working on plant life histories to gather the data necessary for constructing these fitness gain curves by combining data on pollen and seed production, pollen transfer, seedling competition, etc. This is the more interesting because molecular techniques have now made it possible to establish by which individual a seed was fathered, a step crucial for constructing male fitness curves (Snow and Lewis, 1993).

III. Plant Size and Gender in *Cynoglossum officinale*

Cynoglossum officinale is a monocarpic perennial species that we study at the sand dunes of Meijendel (near the Hague, the Netherlands). After germination plants remain vegetative for one or more years and then reproduce and die (de Jong *et al.*, 1990). As in most monocarpic perennials the probability of reproducing in a certain year is not related to the age of the plant but rather is strongly related to plant mass. Only after some threshold size is reached a plant can reproduce. Flowering is restricted to a period of 3 to 4 weeks in May and June. Flowering stalks are 40–90 cm and may bear tens to hundreds of flowers. The red-purple corolla turns to blue before abscission after 2 to 3 days. Cymous inflorescences develop from the axils of alternate stem leaves. Usually one to three flowers are open simultaneously on each cyme. The most common flower visitors are *Bombus* spp.; honeybees are less frequent. The percentage of selfed seeds increases up to about 60% with plant size (Vrieling *et al.*, 1997).

Cynoglossum officinale is a convenient species in which to study reproductive allocation because seeds stay on the plant after seed ripening for 1 to 2 months, which makes it possible to collect individuals after seeds are ripe and determine total number of flowers and total number of seeds at one point in time. As in other members of the Boraginaceae, flowers of *Cynoglossum officinale* have four ovules and may produce up to four seeds. Mostly, however, flowers produce fewer seeds (Fig. 2).

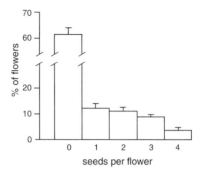

Figure 2   Frequency distribution of the number of seeds per flower in *Cynoglossum officinale*. Data are averaged over 54 plants collected in a population in a coastal dune area.

To study the relationship between vegetative plant mass and seed and flower production we used a simple allometric model (Gould, 1966; Klinkhamer *et al.*, 1990):

$$Y = aX^b \quad \text{or} \quad \log Y = \log a + b \log X,$$

in which X is plant mass and Y is seed number, total seed mass, or flower number.

If $b = 1$, Y is proportional to X.
If $b > 1$, Y increases more than proportionally to X.
If $b < 1$, Y increases less than proportionally to X.

A. Plant Mass and Seed Production

We studied seed and flower production in a number of populations in different years in a coastal dune area. The pattern appeared to be consistent over the years and similar to the one described here for a single population (Klinkhamer and de Jong, 1987). Vegetative plant mass is defined as all mass of the plant except the seeds after drying for 48 hr at 70°C. In the regression of log number of seeds versus log plant mass the regression coefficient is close to 1, indicating that seed number is proportional to plant mass (Fig. 3A). Per gram plant mass, small and large plants produce equal amounts of seeds (Fig. 4A). The same applies to total seed mass per plant (Fig. 3B) because the mass per seed is not related to plant mass.

B. Plant Mass and Flower Production

In the regression of log flower number versus log plant mass the regression coefficient is significantly smaller than 1 (Fig. 3C): flower number increases less than proportionally with plant mass. In other words, per gram plant mass, small plants produce more flowers than larger plants (Fig. 4B).

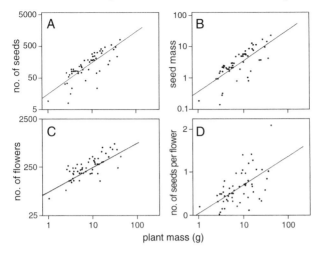

Figure 3 Relationship between plant mass and number of seeds (A), total seed mass (B), number of flowers (C), and number of seeds per flower (D) for a population of *Cynoglossum officinale*. Regression equations are as follows: (A) log $Y = 0.98$ log $X + 1.23$, $R^2 = 0.62$; (B) log $Y = 1.02$ log $X - 0.47$, $R^2 = 0.58$; (C) log $Y = 0.52X + 1.95$, $R^2 = 0.59$; (D) Spearman rank correlation $T = 0.54$. In all cases $P < 0.001$. After Klinkhamer and de Jong (1987).

From the regression it can be estimated that small plants (1 g) produce about 90 flowers per gram plant mass, whereas large plants (60 g) produce only 12 flowers per gram.

C. Plant Mass and Seed-to-Flower Ratio

Only 5% of all flowers produce the maximum number of four seeds, whereas 62% of all flowers do not produce seeds at all (Fig. 2). Mean seed-

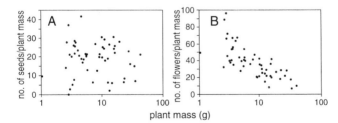

Figure 4 Number of seeds per gram plant mass versus plant mass (A) and number of flowers per gram plant mass versus plant mass (B), calculated from the data presented in Fig. 3. The number of flowers per gram plant decreases significantly with plant mass (see Table I, F-test for the relationship between log number of flowers versus log plant mass); small and large plants produce equal amounts of seeds per gram plant mass (Table I).

to-flower ratio per plant ranges from 0 to 2 (Fig. 3D). Because, per gram vegetative mass, small and large plants produce about equal amounts of seeds and small plants produce more flowers, the seed-to-flower ratio increases with plant mass.

D. Plant Mass and Pollen Production

Based on nine individual flowers sampled per plant, the mean pollen production per flower ranges from 30,000 to 155,000 (P. G. L. Klinkhamer and T. J. de Jong, unpublished, 1993). Variation in number of pollen grains per flower is not related to plant mass: small and large plants produced on average (SD) 54,000 (12,000) pollen grains per flower ($n = 10$ for each group).

E. Plant Mass and Gender

Because the number of pollen grains does not change with plant size, we conclude that small plants emphasize male reproduction (producing relatively many flowers with fewer seeds per flower) and large plants emphasize female reproduction (producing relatively few flowers with more seeds per flower).

IV. Causal Explanations for Increasing Femaleness with Plant Size

There are several causal explanations for the relationship between sex allocation and plant mass (Fig. 5):

1. Large plants have many flowers and attract more pollinators, which results in a higher seed production per flower compared to small plants (Fig. 5A). Because small plants have fewer seeds per flower, flower production is less inhibited by developing seeds, and more resources are spent on the production of flowers.

We found that large plants were more often approached by pollinators, but the proportion of flowers visited after arrival of a bumblebee was smaller on large plants. The net result was that flowers on large plants (100 open flowers) received 50% more visits than flowers on small plants (10 open flowers) (Klinkhamer *et al.*, 1989). Potentially, therefore, this explanation based on pollinator visitation can be true, and as a next step we investigated if seed production was limited by resources or by the level of pollination. In a natural population plants were divided over three treatments (extra hand pollination, extra water added, and extra water and hand pollination) and a control group (de Jong and Klinkhamer, 1989b). In the hand-pollination treatment pollen was supplied to all open flowers three times per week. In the water treatment 6 liters water per week was supplied to a 0.25-cm² surface around the plant. In the dunes where we study *Cynoglossum*

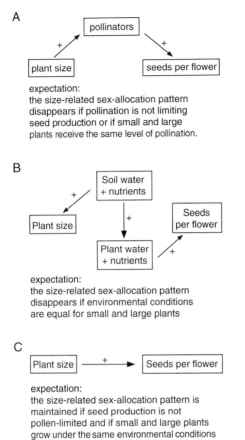

Figure 5 Three possible causal explanations for the relationship between plant mass and number of seeds per flower. (A) Large plants are better pollinated, which results in a higher seed production per flower compared to small plants. (B) Environmental factors such as soil nutrients and soil water enhance both plant size and the condition of the plant as measured by, e.g., internal nutrient concentrations or water potential, and these in turn affect the seed-to-flower ratio. (C) Plant mass directly affects the level of ovule or seed abortion independent of environmental conditions.

officinale, water is a limiting factor for plant growth. Neither the number of seeds per flower nor the number of seeds per plant were increased by experimental pollination, although watering the plants increased both. We found no interaction between the pollination and water treatment. Removing an estimated 50% of the open flowers before they could produce seeds nearly doubled the seed production in the remaining flowers. Therefore, we can conclude that in *Cynoglossum officinale* seed production is limited by resources and not by the level of pollination. We can reject this first explanation for the increase in the number of seeds per flower with plant mass.

2. *Environmental factors such as soil nutrients and soil water enhance both plant size and the condition of the plant as measured by, e.g., internal nutrient concentrations or water potential, and these in turn affect the seed-to-flower ratio (Fig. 5B).*

If this explanation were true we would expect the pattern to disappear if environmental conditions are made equal for small and large plants. To test this idea, we selected 80 plants in the dunes and measured the concentrations of N, P, K, and Mg of a "standard" leaf at the beginning of the formation of the flowering stem (Klinkhamer and de Jong, 1993). Although plants with initially high internal nutrient concentrations indeed bore slightly more seeds, no correlation existed between any of these nutrients and plant mass: large reproducing plants had similar internal nutrient concentrations as small ones. As mentioned before, soil water is a limiting factor for plant growth and reproduction. To make soil water conditions equal for small and large plants we gave 40 of the plants we selected water during the whole period of flower and seed production. The soil surrounding these plants was brought to field capacity three times per week. The remaining 40 plants served as a control group. The water treatment increased average seed production by 50%. However, even when environmental conditions are independent of plant size (the water group), sex allocation appeared to be size dependent (Fig. 6). Moreover, if this explanation were true we would not expect the small plants to produce more flowers per unit of plant mass than large plants as we found in all experiments. We therefore have little evidence in support of this second explanation for the high seed set per flower in large plants.

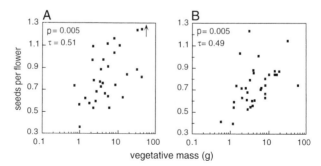

Figure 6 Number of seeds per flower versus plant mass in *Cynoglossum officinale* in a coastal dune area. (A) Water treatment (to make environmental conditions equal for small and large plants). (B) Control. After Klinkhamer and de Jong (1993).

3. Plant mass directly affects the level of ovule or seed abortion independent of environmental conditions (Fig. 5C). Smaller plants have a higher level of abortion, and therefore flower production is less inhibited by developing seeds.

This explanation assumes a trade-off between seed and flower production. In *Cynoglossum* we showed a trade-off between flower and seed production by preventing pollination. Compared to hand-pollinated plants (0.72 seeds per flower), nonpollinated plants produced only 0.18 seeds per flower; at the same time, however, the number of flowers per gram vegetative mass increased from 20 to 45 (Klinkhamer and de Jong, 1987). Also, if half the number of flowers is removed from plants before they can set seed, total flower production (including the removed ones) is increased, and the remaining flowers produce more seeds per flower (de Jong and Klinkhamer, 1989b). Although we do not know through what mechanism plant mass can influence seed abortion, at least two observations are in line with this third explanation. First, the size-dependent allocation to seeds and flowers is maintained when environmental conditions are equal for small and large plants. Second, small plants produce per gram plant mass 3 to 8 times the number of flowers produced by larger plants.

V. Is the Sex Allocation Pattern Found in *Cynoglossum officinale* Representative for Other Hermaphrodite Animal-Pollinated Monocarpic Plants?

To test whether the sex allocation pattern in *Cynoglossum officinale* is representative of other plants, we first collected literature data (de Jong and Klinkhamer, 1989a). These data were of limited use, however. Although there are hundreds of papers on the size-dependent allocation to seeds, only a very few papers also focus on allocation to male reproduction. Apart from our study on *Cynoglossum*, papers on the relationship between plant mass and pollen production seemed to be completely lacking. Although information on flower, fruit, and seed production in relation to plant size is scattered through the literature, we found few papers that combined information for all these aspects of plant reproduction for single species. A common problem is that seed and flower production are not scaled in relation to biomass, but rather with respect to a one-dimensional measure of plant size (e.g., height). The allometric scaling of this measure and biomass produces problems of its own and makes the analysis more difficult (Niklas, 1994). Aside from our work on *Cynoglossum*, we found only one other paper on animal-pollinated hermaphrodite monocarpic plants that gave all necessary raw data on a large number of plants sampled in a natural population. In *Yucca whipplei* (Aker, 1982) results resembled those of *Cynoglossum*: small plants produced relatively many flowers, and the seed-to-flower ratio increased with plant mass. One other paper showed a significant increase of the seed-to-flower ratio with plant mass [*Floerkia proserpinacoides* (Smith, 1983)], although in another paper the relationship was nonsignificantly negative [*Polygonum cascadense* (Hickman, 1975)]. Most papers contained only data on seeds per fruit or fruits per flower. If we also accept these data as indicative for sex allocation, the relative investment in female reproduction increased with plant mass in 28 cases and was not related to plant mass in 1 case, whereas in 5 species a nonsignificant negative correlation was found (de Jong and Klinkhamer, 1989a).

Two other studies on gender and plant mass have been published. Dudash (1991) studied *Sabatia angularis* and found that in large plants more flowers developed into fruits and fruits had twice the number of seeds compared to small plants. Flowers on larger plants produced 1.4 times the number of pollen grains compared to flowers on small plants. Combining these data shows that the ratio of seed to pollen production and thus the relative investment in female reproduction increases with plant mass in *Sabatia angularis*. Damgaard and Loeschcke (1994) studied gender in *Brassica napus* and found that total seed mass divided by the number of flowers varied significantly among genetic lines and increased with aboveground plant mass.

In addition to literature data we collected data on the relationship between plant mass and flower and seed production for another seven monocarpic species, six from Dutch sand dunes and one (*Ipomopsis aggregata*) from the Rocky Mountains. In all species all flowers were counted, and plant mass was determined after seed ripening. In *Ipomopsis aggregata* and *Carlina vulgaris* all seeds were counted. In *Oenothera erythrosepala*, *O. biennis*, *Verbascum thapsus*, *Echium vulgare*, and *Senecio sylvaticus*, the number of seeds per flower on a random selection of flowers, and the number of open flowers, were determined throughout the season. These data together with the total number of open flowers were used to calculate seed production.

The allometric relationship between seed production and plant mass is not consistent among the eight biennial species we studied (Table I). In six of the eight species the coefficient for the regression between log seed number and log plant mass is greater than 1 but only in two cases was this significant. In one of the two species where this coefficient was smaller than 1 this was significantly so. Such variable patterns have also been found by Samson and Werk (1986). For flower production the situation is reversed to that of seed production (Table I). In six of eight species the allometric coefficient was smaller than 1, and in three species this was significant. In one of the two species where this coefficient was greater than 1 this was significant. More consistent was the pattern of relative allocation to seeds and flowers. In 6 of 8 species we studied the relative allocation to female reproduction (as indicated by the seed to flower ratio) increased signifi-

Table I   Relationship between Vegetative Plant Mass and Number of Seeds per Flower, Number of Flowers, and Number of Seeds[a]

Species	Seeds per flower versus plant mass T	N	Range
Oenothera erythrosepala	0.52**	29	105–239
Oenothera biennis	0.67***	29	116–226
Cynoglossum officinale[b]	0.54***	54	0–2.11
Ipomopsis aggregata	0.65***	50	2.2–9.8
Senecio sylvaticus	0.54***	29	0.74–0.97
Verbascum thapsus	0.61**	19	118–710
Echium vulgare[c]	0.30	28	0.4–1.6
Carlina vulgaris	−0.15	38	0.57–0.97

Species	Log number of flowers versus log plant mass Intercept	Slope	R^2	F
Oenothera erythrosepala	0.48	0.96	0.93	0.58
Oenothera biennis	0.94	0.70	0.90	48***
Cynoglossum officinale[b]	1.96	0.51	0.59	70***
Ipomopsis aggregata	1.74	0.87	0.87	7.0**
Senecio sylvaticus	3.48	0.93	0.94	2.8
Verbascum thapsus	0.84	0.91	0.81	0.73
Echium vulgare[c]	1.86	1.09	0.96	4.2*
Carlina vulgaris	1.84	1.06	0.67	0.28

Species	Log number of seeds versus log plant mass Intercept	Slope	R^2	F
Oenothera erythrosepala	2.51	1.07	0.92	1.2
Oenothera biennis	3.02	0.82	0.92	16.1***
Cynoglossum officinale[b]	1.16	1.04	0.53	0.23
Ipomopsis aggregata	2.45	1.20	0.87	9.36***
Senecio sylvaticus	3.45	0.95	0.95	1.33
Verbascum thapsus	2.64	1.32	0.72	2.67
Echium vulgare[c]	1.72	1.17	0.95	10.4*
Carlina vulgaris	1.79	1.04	0.64	0.09

[a] For seeds per flower we used Spearman rank correlations. For flowers and seeds we used a linear regression of log number of flowers or log number of seeds versus log plant mass. We also tested if in these regressions the regression coefficient was significantly different from 1 (F-test). A regression coefficient smaller than 1 means that, e.g., number of flowers increases less than proportionally with plant mass; a coefficient larger than 1 means that the increase is more than proportional. Significant test results are indicated as follows: *, $P < 0.05$; **, $P < 0.01$; ***, $P < 0.005$. All species were studied in Meijendel, a coastal dune area in the Netherlands, except *I. aggregata* which was studied at Crested Butte (Colorado, USA).

[b] Data from Klinkhamer and de Jong (1987).

[c] Data from Klinkhamer *et al.* (1994b).

cantly with plant mass (Table I). In the other two the relationship was not significant. The species in which the relative alocation to female reproduction increased with plant mass were also the species in which small plants produced more flowers per gram plant mass than large plants. Pollen production was studied only in *Echium vulgare* and appeared to be similar in small and large plants (Klinkhamer *et al.*, 1994b). The conclusion from both our own studies and the literature data is unequivocal: in monocarpic animal-pollinated plants, small individuals emphasize male reproduction and large individuals emphasize female reproduction.

VI. Is Increased Emphasis on Female Reproduction with Plant Size in Insect-Pollinated Plants Adaptive? A Comparison between Animal- and Wind-Pollinated Plants

Although there appears to be considerable variation in the allometric relationship between seed and flower production and plant mass among species, we concluded that small plants emphasize male reproduction and large plants emphasize female reproduction. It should be stressed, however, that we only studied monocarpic species and that at least for some perennial species other allocation patterns have been reported (Willson and Rathcke, 1974). The pattern we found for animal-pollinated, monocarpic species is consistent with the idea that the male fitness curve levels off more quickly with plant size than the female fitness curve and that therefore fitness of large plants is increased mostly by emphasizing female reproduction and fitness of small plants is increased most by emphasizing male reproduction (Fig. 1, right-hand side). Models based on a decelerating fitness gain curve for male and a linear one for female function predict that small plants should be only male whereas large plants should be simultaneously male and female (e.g., de Jong and Klinkhamer, 1989a). Strictly male plants are, however, not often found. In the species we studied and in the literature data (a total of 43 species) individuals without any seed production were found in only two cases. Interestingly, data collected for sequential hermaphrodites (with alternate reproduction as male or female by the same individual) support the hypothesis that small individuals allocate to male function and that allocation is shifted to female function as the size of an individual exceeds a certain threshold. Allocation can be reversed from female to male under stressful growth conditions (see Freeman *et al.*, 1980, and references in Niklas, 1994).

If the size-dependent sex allocation pattern results from sexual selection as we depicted, we can make several predictions:

1. If local resource competition is low (e.g., in species with good seed dispersal mechanisms), the female fitness curve does not level off and we expect a strong relationship between femaleness and plant size.

2. In self-incompatible species the effects of geitonogamy are more severe, and we expect the male fitness curve to level off more strongly than in self-compatible species. Consequently, we expect a stronger relationship between femaleness and plant size in self-incompatible species than in self-compatible species

3. If the level of geitonogamy is independent of plant size, we expect that the size-dependent pattern of sex allocation disappears or is reversed; then, large plants should emphasize male reproduction, and small plants should emphasize female reproduction.

Unfortunately, the first two hypotheses cannot be tested with the available data, but they may guide future research. The third hypothesis can be tested by comparing wind- and animal-pollinated species. In wind-pollinated species the male fitness curve is expected to be linear (Charnov, 1982; Charlesworth and Charlesworth, 1981; Schoen and Stewart, 1986), or to accelerate (Burd and Allen, 1988). Therefore, we would expect the female gain curve to level off more quickly than the male curve, and if plants can adapt their allocation patterns we would expect the relative allocation to female reproduction to decrease with plant mass in wind-pollinated species (Fig. 7). In contrast, we expect femaleness to increase with plant mass in animal-pollinated species. Unfortunately, the available data for hermaphrodite species include only two wind-pollinated species, namely, *Vulpia fasiculata* and *Beta maritima,* and in both femaleness increases with plant mass (Watkinson, 1982; Boutin-Stadler, 1987). Interestingly, some evidence for the third hypothesis comes from data on size-dependent sex allocation in monoecious plants (Freeman *et al.*, 1981; Bickell and Freeman, 1993). Of a total of 23 species, the 8 species that are entomophilous increased

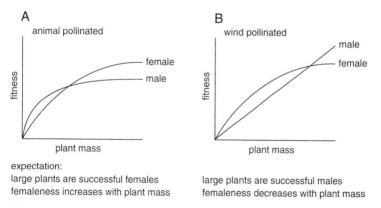

Figure 7   The expected curves describing the relationship between fitness and plant mass for animal- and wind-pollinated species.

Table II Distribution of Postitive and Negative Relationships between the Ratio of Investment in Female and Male Reproduction and Plant Size in Animal- and Wind-Pollinated Monoecious Plants[a]

	Sign of relationship	
Mode of pollination	+	−
Animal	8	0
Wind	6	9

[a] $P < 0.01$, Fisher Exact Probability Test. Data from Freeman *et al.* (1981) and data reviewed in Bickell and Freeman (1993).

femaleness when large, whereas 9 of 15 anemophilous species increased maleness when large (Table II).

In conclusion,

1. Sex allocation in plants is size dependent.

2. Size-dependent sex allocation patterns are different for groups of plants with different pollination systems.

3. These patterns seem to be adaptive, but much more knowledge is needed about the relationship between plant size and fitness.

4. There is need for a better understanding of the physiological regulation of flower, pollen, and seed production and abortion of ovules and seeds.

References

Aizen, M. A., Searcy, K. B., and Mulcahy, D. L. (1990). Among- and within-flower comparisons of pollen tube growth following self- and cross-pollinations in *Dianthus chinensis* (Caryophyllaceae). *Am. J. Bot.* **77,** 671–676.

Aker, C. L. (1982). Regulation of flower, fruit and seed production by a monocarpic perennial, *Yucca whipplei. J. Ecol.* **70,** 357–372.

Ashman, T. L., and Baker, I. (1992). Variation in floral sex allocation with time of season and currency. *Ecology* **73,** 1237–1243.

Ayre, D. J., and Whelan, R. J. (1989). Factors controlling fruit set in hermaphroditic plants: Studies with the Australian Proteaceae. *Trends Ecol. Evol.* **4,** 267–272.

Bickell, A. M., and Freeman, D. C. (1993). Effects of pollen vector and plant geometry on floral sex ratio in monoecious plants. *Am. Midl. Nat.* **130,** 239–247.

Boutin-Stadler, V. (1987). Selection sexuelle et dynamique de la sterilite male dans les populations de betteraves sauvage, *Beta maritima* L. These de L'univerisite des sciences et technique de Lille Flandre Artois, France.

Brunet, J. (1992). Sex allocation in hermaphroditic plants. *Trends Ecol. Evol.* **7,** 79–84.

Burd, M. (1994). Bateman's principle and plant reproduction: The role of pollen limitation in fruit and seed set. *Bot. Rev.* **60**, 83–139.

Burd, M., and Allen, T. F. H. (1988). Sexual allocation strategy in wind-pollinated plants. *Evolution* **42**, 403–407.

Charlesworth, D., and Charlesworth, B. (1981). Allocation of resources to male and female functions in hermaphrodites. *Biol. J. Linn. Soc.* **15**, 57–74.

Charnov, E. L. (1982). "The Theory of Sex Allocation." Princeton Univ. Press, Princeton, New Jersey.

Charnov, E. L. (1984). Behavioural ecology of plants. *In* "Behavioural Ecology" (J. R. Krebs and N. B. Davis, eds.), 2nd Ed., pp. 362–380. Blackwell, Oxford.

Charnov, E. L., and Bull, J. J. (1977). When is sex environmentally determined? *Nature (London)* **266**, 828–830.

Charnov, E. L., and Bull, J. J. (1985). Sex allocation in a patchy environment: A marginal value theorem. *J. Theor. Biol.* **115**, 619–624.

Damgaard, C., and Loeschcke, V. (1994). Genotypic variation for reproductive characters, and the influence of pollen–ovule ratio on selfing rate in rape seed (*Brassica napus*). *J. Evol. Biol.* **7**, 599–607.

de Jong, T. J., and Klinkhamer, P. G. L. (1989a). Size-dependency of sex-allocation in hermaphroditic, monocarpic plants. *Funct. Ecol.* **3**, 201–206.

de Jong, T. J., and Klinkhamer, P. G. L. (1989b). Limiting factors for seed production in *Cynoglossum officinale*. *Oecologia* **80**, 167–172.

de Jong, T. J., and Klinkhamer, P. G. L. (1994). Plant size and reproductive success through female and male function. *J. Ecol.* **82**, 399–402.

de Jong, T. J., Klinkhamer, P. G. L., and Boorman, L. A. (1990). *Cynoglossum officinale*. Biological Flora of the British Isles, No. 170. *J. Ecol.* **78**, 1123–1144.

de Jong, T. J., Klinkhamer, P. G. L., and van Staalduinen, M. J. (1992). The consequences of pollination biology for mass blooming or extended blooming. *Funct. Ecol.* **6**, 606–615.

de Jong, T. J., Waser, N. M., and Klinkhamer, P. G. L. (1993). Geitonogamy: The neglected side of selfing. *Trends Ecol. Evol.* **8**, 321–325.

de Jong, T. J., Klinkhamer, P. G. L., and Rademaker, M. (1997). How geitonogamous selfing affects sex allocation in hermaphrodite plants. Submitted for publication.

Dudash, M. R. (1991). Plant size effects on female and male function in hermaphroditic *Sabiata angularis* (Gentianaceae). *Ecology* **72**, 1004–1012.

Freeman, D. C., Harper, K. T., and Charnov, E. L. (1980). Sex change in plants: Old and new observations and new hypothesis. *Oecologia* **47**, 222–232.

Freeman, D. C., Mcarthur, E. D., Harper, K. T., and Blauer, A. C. (1981). Influence of environment on the floral sex ratio of monoecious plants. *Evolution* **35**, 194–197.

Geber, M. A. (1985). The relationship of plant size to self-pollination in *Mertensia ciliata*. *Ecology* **66**, 762–772.

Goldman, D. A., and Willson, M. F. (1986). Sex allocation in functionally hermaphroditic plants: A review and critique. *Bot. Rev.* **52**, 157–194.

Gould, S. J. (1966). Allometry and size in ontogeny and phylogeny. *Biol. Rev.* **41**, 587–640.

Hickman, J. C. (1975). Energy allocation and niche differentiation in four co-existing annual species of *Polygonum* in western North America. *J. Ecol.* **63**, 689–701.

Horovitz, A. (1978). Is the hermaphrodite flowering plant equisexual? *Am. J. Bot.* **65**, 485–486.

Iwasa, Y. (1990). Evolution of the selfing rate and resource allocation models. *Plant Species Biology* **5**, 19–30.

Klinkhamer, P. G. L., and de Jong, T. J. (1987). Plant size and seed production in the monocarpic perennial *Cynoglossum officinale*. *New Phytol.* **106**, 773–783.

Klinkhamer, P. G. L., and de Jong, T. J. (1990). Effects of plant size, plant density and sex differential nectar reward on pollinator visitation in the protandrous *Echium vulgare* (Boraginaceae). *Oikos* **57**, 399–405.

Klinkhamer, P. G. L., and de Jong, T. J. (1993). Phenotypic gender in plants: Effects of plant size and environment on allocation to seeds and flowers in *Cynoglossum officinale*. *Oikos* **67**, 81–86.

Klinkhamer, P. G. L., de Jong, T. J., and de Bruyn, G. J. (1989). Plant size and pollinator visitation in *Cynoglossum officinale*. *Oikos* **54**, 201–204.

Klinkhamer, P. G. L., de Jong, T. J., and Meelis, E. (1990). How to test for size-dependency of reproductive effort. *Am. Nat.* **135**, 291–300.

Klinkhamer, P. G. L., de Jong, T. J., and Metz, J. A. J. (1994a). Why plants can be too attractive; a discussion of measures to estimate male fitness. *J. Ecol.* **82**, 191–194

Klinkhamer, P. G. L., de Jong, T. J., and Nell, H. W. (1994b). Limiting factors for seed production and phenotypic gender in the gynodioecious species *Echium vulgare* (Boraginaceae). *Oikos* **71**, 469–478.

Lloyd, D. G. (1984). Gender allocation in outcrossing cosexual plants. *In* "Perspectives on Plant Population Ecology" (R. Dirzo and J. Sarukhan eds.), pp. 277–300. Sinauer, Sunderland, Massachusetts.

Lloyd, D. G. (1988). Benefits and costs of biparental and uniparental reproduction in plants. *In* "The Evolution of Sex" (R. E. Michod and B. R. Levin, eds.), pp. 233–252. Sinauer, Sunderland, Massachusetts.

Lloyd, D. G., and Bawa, K. S. (1984). Modification of gender of seed plants in varying conditions. *Evol. Biol.* **17**, 255–338.

Morgan, M. (1993). Fruit to flower ratios and trade-offs in size and number. *Evol. Ecol.* **7**, 219–232.

Niklas, K. J. (1994). "Plant Allometry. The Scaling of Form and Process." Univ. of Chicago Press, Chicago.

Pleasants, J. M. (1991). Evidence for short-distance dispersal of pollinia in *Asclepias syriaca* L. *Funct. Ecol.* **5**, 75–82.

Robertson, A. W. (1992). The relationship between floral display size, pollen carryover and geitonogamy in *Myosotis colensi* (Kirk) Macbride (Boraginaceae). *Biol. J. Linn. Soc.* **46**, 333–349.

Roff, D. A. (1992). "The Evolution of Life Histories: Theory and Analysis." Chapman & Hall, New York.

Salisbury, E. J. (1942). "The Reproductive Capacity of Plants." Bell, London.

Samson, D. A., and Werk, K. S. (1986). Size-dependent effects in the analysis of reproductive effort in plants. *Am. Nat.* **127**, 667–680.

Saumitou-Laprade, P., Cuguen, J., and Vernet, P. (1994). Cytoplasmic male sterility in plants: Molecular evidence and the nucleocytoplasmic conflict. *Trends Ecol. Evol.* **9**, 431–435.

Schoen, D. G., and Stewart, S. C. (1986). Variation in male reproductive investment and male reproductive success in white spruce. *Evolution* **40**, 1109–1120.

Smith, B. H. (1983). Demography of *Floerkia proserpinacoides*, a forest-floor annual. II. Density-dependent reproduction. *J. Ecol.* **71**, 405–412.

Snow, A. A., and Lewis, P. O. (1993). Reproductive traits and male fertility in plants. *Annu. Rev. Ecol. Syst.* **24**, 331–351.

Spalik, K. (1991). On evolution of andromonoecy and 'overproduction' of flowers: A resource allocation model. *Biol. J. Linn. Soc.* **42**, 325–336.

Stanton, M. L., and Galloway, L. F. (1990). Natural selection and allocation to reproduction in flowering plants. *Lect. Math. Life Sci.* **22**, 1–49.

Stearns, S. C. (1992). "The Evolution of Life Histories." Oxford Univ. Press, Oxford.

Stephenson, A. G. (1981). Flower and fruit abortion: Proximate causes and ultimate function. *Annu. Rev. Ecol. Syst.* **12**, 253–279.

Sutherland, S. (1986a). Floral sex ratios, fruit-set and resource allocation in plants. *Ecology* **67**, 991–1001.

Sutherland, S. (1986b). Patterns of fruit-set: What controls fruit–flower ratios in plants? *Evolution* **40,** 117–128.

Sutherland, S., and Delph, L. F. (1984). On the importance of male-fitness in plants: Patterns of fruit-set. *Ecology* **65,** 1093–1104.

Thomson, J. D., and Brunet, J. (1990). Hypothesis for the evolution of dioecy in seed plants. *Trends Ecol. Evol.* **5,** 11–16.

Vrieling, K., Saumitou-Laprade, P., Cuguen, J., van Dyk, H., de Jong, T. J., and Klinkhamer, P. G. L. (1997). Pollinator behaviour and geitonogamous selfing in relation to plant size in *Cynoglossum officinale* L. Submitted for publication.

Watkinson, A. (1982). Factors affecting the density response of *Vulpia fasciculata. J. Ecol.* **70,** 149–161.

Willson, M. F., and Burley, N. (1983). "Mate Choice in Plants: Tactics, Mechanisms and Consequences." Princeton Univ. Press, Princeton, New Jersey.

Willson, M. F., and Rathcke, B. J. (1974). Adaptive design of the floral display of *Asclepias syriaca* L. *Am. Midl. Nat.* **92,** 47–57.

Yampolsky, E., and Yampolsky, H. (1922). Distribution of sex forms in phanerogamic flora. *Bibl. Genet.* **3,** 1–62.

10

Allocation, Leaf Display, and Growth in Fluctuating Light Environments

David Ackerly

I. Introduction

Plants, and other organisms, live in highly variable environments. The effect of this variability on the growth and performance of individual plants depends critically on the scale of environmental variation and on the temporal and spatial scale of plant responses. For example, transient sun patches in the understory of a forest provide bursts of high energy that are critical for carbon gain in these environments. The utilization of this energy depends on the duration of the patch because various physiological processes (e.g., instantaneous photosynthesis, stomatal opening, enzyme activity) respond at different rates to the change in environment. Additionally, for any particular process, such as photosynthetic rates or allocational patterns, species differ in their rate of response to environmental changes, owing to variation in underlying physiological and developmental mechanisms of response. The responses of different traits to a change in the environment, which comprise a subset of the overall phenotypic plasticity of an individual, have a strong influence, in an ecological and evolutionary context, on how the scale of environmental fluctuations influences whole plant performance (i.e., growth, survival, and reproductive output; see Caldwell and Pearcy, 1994). The effects of variable light environments are best understood in the case of sunflecks, in which the duration and frequency of light patches affects total assimilation via responses in an array of physiological processes (Pearcy *et al.*, 1994). Whole plant carbon gain is also influenced by the

Plant Resource Allocation

231

Copyright © 1997 by Academic Press.
All rights of reproduction in any form reserved.

temporal scale of diurnal fluctuations in light environments, even when the total amount of light received is kept constant (Garner and Allard, 1931; Hogan, 1990; Sims and Pearcy, 1993; Wayne and Bazzaz, 1993). Physiological and allocational responses to environmental fluctuations on longer time scales (i.e., exceeding 1 day) have received less attention (Pearcy and Sims, 1994), particularly the response to fluctuating light environments, rather than discrete changes between two sets of conditions. These are important components of plant response to gap creation and closure in forests, and short- and long-term climatic fluctuations (including annual seasonality; Mitchell, 1976). In this chapter, I discuss results of two studies on the response of tree seedlings to changing light environments, focusing on the developmental mechanisms and time scale of response of allocation and leaf area display, and on the significance of these responses for whole plant performance.

Problems of scale and scaling have attracted considerable attention among ecologists. Various analytical techniques, such as autocorrelation and fractal analysis, have been employed to describe the structure and scale of spatial and temporal environmental variation (e.g., Robertson and Gross, 1994). To understand the effects of this heterogeneity on plant performance, it is important to focus on the scale of this variation relative to the scale of plant responses. The importance of relative scale is recognized in the concept of "environmental grain," introduced into population biology by MacArthur and Levins (1964; Levins, 1968). At the level of the individual organism, coarse-grained environments are those in which the "patches" last longer than the life span of the plant, so the individual encounters only one of the various conditions during its lifetime. Fine-grained environments, on the other hand, are those in which patches are relatively short such that each individual encounters a succession of different conditions during its life cycle. (In this chapter I focus on temporal heterogeneity; analogous definitions can be given for spatial variation.) The expression and significance of phenotypic plasticity depends critically on environmental grain. In coarse-grained environments, each individual manifests only one of the potential phenotypes for a given trait. For example, in a population occupying a coarse-grained mosaic of soil types, individuals in low-nutrient soils might exhibit greater root:shoot ratios compared to those in high-nutrient patches. This plasticity in allocational patterns would be evident at the population level, expressed in the variation among individuals. In a fine-grained environment, however, each individual may be able to express a wide range of the possible phenotypes during its lifetime. For example, annual fluctuations in rainfall in a population of perennials might elicit considerable year-to-year variation in partitioning of new growth to roots and shoots. In this case, phenotypic variation due to environment would be evident within rather than between individuals, and plasticity can be apparent at the individual and population level.

In fine-grained environments, the expression of plasticity by individuals depends further on the relative scale of environmental variation and phenotypic responses (Gross, 1986; Kuiper and Kuiper, 1988). As introduced above for the example of sunflecks, a transient change in the light environment will only elicit a physiological change in traits with relatively rapid response times, such as photosynthesis or stomatal behavior. Other traits, such as biomass partitioning, are essentially buffered from these fine-scale fluctuations by their slower responses. Effectively, the response time of a trait acts as a filter on the perception of environmental fluctuations (see Allen and Starr, 1982). If environmental fluctuations are rapid, relative to response time, then the trait will exhibit an average value with little temporal variation. If fluctuations are slower than the characteristic response time, however, then the trait may track these fluctuations and possibly exhibit adaptive adjustments in each patch that enhance the physiological function and performance of the individual.

Response times of different traits vary from fractions of a second, for molecular transformations of phytochrome, to months or years, for alterations in biomass distribution and canopy structure (Fig. 1; modified from Gross, 1986, and Chazdon, 1988). This range in response times reflects the spectrum of physiological and developmental mechanisms underlying different traits. For traits expressed at the whole plant level (e.g., mean leaf properties, allocation, canopy structure), changes in phenotypic characteristics occur through two distinct mechanisms: (1) physiological and anatomical changes within existing tissues and (2) production of new tissues that exhibit distinct physiological, anatomical, or morphological characteristics in response to the current ambient environment (Bradshaw, 1965). The relative importance of these mechanisms depends critically on the life span and rate of turnover of different tissues and organs (Grime and Campbell, 1991). Physiological acclimation of existing tissues is expected in species with slow growth rates, as mature tissues must function in a range of conditions, and some of these changes may be rapid and reversible within the organ, depending on the particular biochemical and physiological mechanisms of acclimation (see Thornley, 1991). For traits that are adjustable within mature tissues, the rate of response at the whole plant level will reflect the dynamics of responses at the cell and tissue level.

In contrast, for traits that are not flexible in mature tissues, responses at the canopy or whole plant level following an environmental change will depend on production of new organs and loss of tissues formed in the previous environment. The rate of change in proportional biomass allocation, following a change in the partitioning of new growth, is closely tied to plant relative growth rate (Ackerly, 1993; see Appendix). Responses of

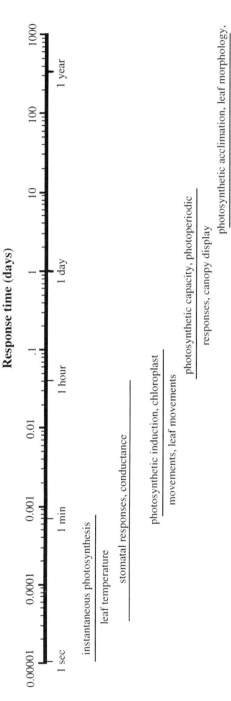

Figure 1 Response times of physiological, developmental, and morphological traits of plants following a change in environmental conditions. Adapted from Gross (1986) and Chazdon (1988).

canopy level characters will also be particularly rapid in fast-growing plants, as high relative growth rates are correlated with short leaf life span and rapid leaf turnover in the canopy (Reich *et al.*, 1992; Ackerly, 1996). Rapid leaf turnover may also be precipitated by changes in light environments, particularly sudden exposure to high light which may cause a plant to drop all of its leaves and produce a new cohort with appropriate adjustments for higher light conditions (e.g., Bazzaz and Carlson, 1982). For a trait that exhibits distinct values in leaves produced in low- and high-light environments, the whole plant response can be expressed as the weighted mean of the values on leaves produced in each environment. Leaf life span will determine the overall duration of the whole plant response, as a complete adjustment to the new conditions will not be observed until all leaves produced in the original environment have died. If a plant is growing rapidly, and total leaf area is increasing, the proportion of standing leaf area that has been produced since an environmental change will increase rapidly, and then slowly approach an asymptote of 100% as the older leaves die. For example, if leaf area is increasing at a relative growth rate of 0.05 day^{-1}, and leaf life span is 100 days, then within 15 days after a change in the environment, 50% of the standing leaf area will have been produced since the change. Thus, rates of change in whole plant allocation and canopy level characters will be strongly associated with plant growth rates, and turnover of organs may be a particularly important mechanism of whole plant response for rapidly growing plants (see Grime and Campbell, 1991).

Experimental studies of plant responses following changes in light environment have focused on physiological and anatomical responses at the leaf level, and allocational responses of the whole plant. Studies of leaf level responses demonstrate that changes in some physiological and anatomical traits can be detected over a period of one to several days following a change in light environment (e.g., Jurik *et al.*, 1979; Sims and Pearcy, 1991), but these changes are generally lower in magnitude than those that occur in new leaves that emerge following the transfer (e.g., Osmond *et al.*, 1988; Sims and Pearcy, 1992). On a longer time scale, leaves produced in different seasons in temperate and tropical forests can exhibit physiological differences when they are measured at any particular point in time (e.g., Shultz *et al.*, 1989; Koike *et al.*, 1992), but the relative importance of environmental differences during leaf development versus differences in leaf age has not been determined. Tropical pioneer species may completely replace the leaves in their evergreen canopies three or four times per year, and there is some evidence that the physiological characteristics of these leaves track seasonal changes in environmental conditions (Ackerly, 1993). "Switching" experiments (involving transfers of plants between light levels) conducted on fast-growing herbaceous plants demonstrate the potential for rapid responses in whole plant allocation. For example, in *Impatiens parviflora* transferred from 42 to 7% light availability, leaf area ratio of individuals doubled in less than a week, though it was still less than that of control plants (Evans and Hughes, 1961; see also Rice and Bazzaz, 1989a). Similar experiments have been conducted on tree seedlings, particularly of tropical species, to examine potential responses to the sudden changes in light that accompany canopy disturbance (Fetcher *et al.*, 1983, 1987; Oberbauer and Strain, 1985; Bongers *et al.*, 1988; Osunkoya and Ash, 1991; Popma and Bongers, 1991; Strauss-Debenedetti and Bazzaz, 1991). Allocation and growth responses in these experiments were found to depend on both the magnitude of the environmental change and on the species studied. The adjustment of the switched plants to the new environment is generally more rapid and complete in transfers among higher light levels and in faster growing species (e.g., Fetcher *et al.*, 1983).

These studies demonstrate the potential for significant changes in allocation and leaf area display at the whole plant level in response to temporal changes in environmental conditions. However, few studies have evaluated the responses of these traits in relation to underlying developmental mechanisms, such as leaf turnover rates, and addressed both short-term dynamics and longer term responses. In addition, responses of traits at the whole plant level to slowly oscillating light environments, rather than simply a single environmental change, have received little attention. Oscillating fluctuations in the environment on time scales greater than 1 day occur due to passing weather systems, seasonal and interannual climatic changes (Mitchell, 1976), and disturbance and recovery of vegetation. The effective scale of these oscillations depends on the life span of the plants; shade-tolerant tree saplings and understory species may experience several disturbance events (i.e., treefalls) during their lifetime (Canham, 1985), and the extent of physiological adjustment to the abrupt increase in light and the subsequent return to shadier conditions has received little attention. In this chapter I present the results of two experiments that address these issues, focusing on the following questions:

1. How do the differences in response times of different traits expressed at the whole plant level relate to the physiological and developmental mechanisms of trait expression?

2. Do phenotypic traits exhibit "integrating" and "tracking" responses to environmental fluctuations that are shorter and longer, respectively, than the characteristic response time of the trait?

3. In a slowly fluctuating environment, does the time scale of the oscillations influence total biomass gain, independently of the total amount of light received?

II. Rate and Pattern of Response Following an Environmental Change

The first experiment presented here involves a reciprocal transfer of tropical tree seedlings between sun and shade environments, followed by repeated harvests to determine the time course of responses following the change. The objectives of this experiment were (i) to assess the pattern and rate of response of different traits, expressed at the whole plant level, in relation to the mechanisms underlying trait expression; (ii) to examine the relative roles of individual leaf response versus leaf turnover as mechanisms of response at the canopy level; and (iii) to predict, on the basis of responses to a single transfer, the response patterns that would be observed in a slowly oscillating environment; these predictions were then tested in an independent experiment, presented in Section III. A more detailed description of the methods and results of this experiment can be found in Ackerly (1993).

A. Experiment I: Reciprocal Transfer between Light Environments

This experiment was conducted on seedlings of two tropical pioneer tree species, *Cecropia obtusifolia* and *Heliocarpus appendiculatus,* grown from seed collected at Los Tuxtlas Tropical Biology Station, Veracruz, Mexico. The seedlings were initially placed in one of two environments: high light ($\sim$12 mol m^{-2} day^{-1}) or low light ($\sim$6 mol m^{-2} day^{-1}), created using a blue acetate filter that reduced light quantity and altered the red : far red (R : FR) ratio appropriately. These light levels are in the range of those observed in medium to large gaps in the forest at Los Tuxtlas (Ackerly and Bazzaz, 1995). After 10 weeks, *Cecropia* and *Heliocarpus* seedlings had produced 13 and 20 leaves and were approximately 30 and 40 cm tall, respectively. At this time, half of the individuals in each light environment were transferred to the contrasting environment, and the other half remained in the initial conditions, creating four experimental treatments. The treatments are designated by the pre- and postswitch light levels: LL and HH refer to the low and high light controls, LH is the switch from low to high light, and HL is the switch from high to low light. Detailed records of leaf production and loss were maintained for each plant following the transfer. Starting on the day of the transfer, six and eight harvests, for *Cecropia* and *Heliocarpus* respectively, were conducted over an 80-day period ($n = 8$ plants/treatment/harvest; harvests were on days 0, 5, 10, 20, 41, and 80 for *Cecropia* and days 0, 3, 6, 10, 17, 29, 50, and 80 for *Heliocarpus*). At each harvest, the height, basal diameter, main stem node number, number of standing leaves, length and area of each main stem leaf, and total area of branch leaves were measured. Plants were then divided into roots, main stem, branches, branch leaves, main stem leaf petioles, and individual main stem leaves for determination of dry biomass and proportional allocation.

At the time of the environmental transfer, the seedlings of these species exhibited a variety of typical sun/shade responses. In both species, the plants in high light exhibited greater total biomass, lower allocation to stems and petioles, higher allocation to roots, lower ratios of leaf area to root biomass, and lower specific leaf area and leaf area ratio. Interestingly, proportional allocation to leaf biomass was identical in the two light levels, but high-light plants had equal or lower total leaf area, despite their greater total biomass, due to lower specific leaf area. Many tropical pioneer tree species, including the two studied here, exhibit continuous leaf birth and death at approximately equal rates, leading to continual canopy turnover and more or less constant leaf numbers on each shoot (Ackerly, 1996). In *Heliocarpus*, a mean of 16 to 18 leaves were maintained on the main stem of each seedling across all four treatments, and leaf life span averaged 53 to 60 days. Owing to increases in the size of successive leaves, 50% turnover of leaf area occurred within 20 days following the experimental transfers (Fig. 2a). *Cecropia* maintained an average of 10 leaves on the stem, with slower leaf birth and death rates and a leaf life span of 88 to 100 days; 50% turnover of leaf area was observed in 35 to 40 days (Fig. 2b). On the basis of these differences in leaf turnover rates, it was expected that *Heliocarpus* would exhibit more rapid responses of canopy level traits following a change in the environment. For traits that only exhibit responses in newly produced leaves, complete adjustment to the new environment would not be expected for 60 days in *Heliocarpus* and 100 days in *Cecropia*, reflecting leaf life span.

The most dramatic responses following the change in light environment were observed in specific leaf area and leaf area ratio (which respond in parallel since leaf weight ratio shows little variation among treatments). Significant effects of the new environment on specific leaf area were observed in *Heliocarpus* within 3 days, whereas effects of the preswitch environment persisted only until the harvest 17 days after the transfer (Table I). In contrast, effects of the new, postswitch environment were not detected in *Cecropia* until day 10, and preswitch effects persisted until 41 days after the transfer. Similar patterns were observed for leaf area ratio. Thus, as expected, the species with more rapid canopy turnover also exhibited more rapid response to the new environment for these canopy level traits. However, these responses were much more rapid than expected solely on the basis of leaf turnover rates, suggesting that changes in leaves produced before the switch played an important role. Examination of specific leaf area (SLA) 10 days after the switch in *Heliocarpus* supports this suggestion (Fig. 3). In new leaves expanded following the switch, SLA was determined by the postswitch environment. In intermediate leaves, which at the time of the switch were 20 days old and fully expanded, SLA in the LH and HL

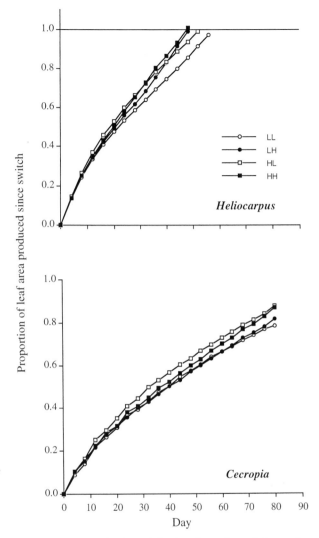

Figure 2   Accumulation of new leaf area following the reciprocal change between low and high light levels in (top) *Heliocarpus appendiculatus* and (bottom) *Cecropia obtusifolia*. The number of leaves per plant was fairly constant over this time, but total leaf area increased because of increases in the size of individual leaves. As a result, 50% of the standing leaf area had been produced after the switch within 20 and 40 days in *Heliocarpus* and *Cecropia*, respectively.

treatments was intermediate between the values observed in the low- and high-light controls. Old leaves, in contrast, were no longer responsive to the change, and their SLA reflects only the preswitch environment.

Table I   Results of Analysis of Variance for Whole Plant Responses following a Reciprocal Transfer between High- and Low-Light Environments[a]

Species	Trait	Factor	Day 0	5	10	20	41	80
Cecropia	Specific leaf area	Pre	***	***	**	**	***	ns
		Post	ns	ns	*	***	***	***
	Leaf area ratio	Pre	***	***	*	ns	ns	ns
		Post	ns	ns	ns	***	***	*
	Support weight ratio	Pre	***	***	*	**	**	ns
		Post	ns	ns	ns	ns	*	*
	Height	Pre	***	***	**	**	*	ns
		Post	ns	ns	ns	ns	*	**
	Height extension rate	Pre		ns	ns	ns	ns	ns
		Post		*	ns	***	***	***

			Day 0	3	6	10	17	29	50	80
Heliocarpus	Specific leaf area	Pre	***	***	*	ns	*	ns	ns	ns
		Post	ns	*	***	**	***	***	***	**
	Leaf area ratio	Pre	***	***	*	ns	ns	ns	ns	ns
		Post	ns	ns	~	~	**	***	**	~
	Support weight ratio	Pre	***	**	*	*	*	*	~	*
		Post	ns	ns	*	*	~	*	ns	*
	Height	Pre	~	~	ns	*	ns	ns	~	ns
		Post	ns	ns	ns	ns	ns	ns	**	*
	Height extension rate	Pre		**	ns	ns	ns	ns	ns	ns
		Post		***	**	~	ns	***	**	***

[a] From Ackerly (1993). Two-factorial model I ANOVA was conducted independently at each harvest to test for pre- and postswitch effects, in order to determine how long effects of the initial environment persist following the transfer, and how much time is required for effects of the new environment to become apparent. There were no significant interactions between pre- and postswitch environments, and this term is omitted from the table. Results for height extension rate are offset in the table because this trait is measured in the interval between harvests, not on the day of a harvest. ns, Not significant; ~, $p < 0.1$; *, $p < 0.05$; **, $p < 0.01$; ***, $p < 0.001$.

The rapid responses of whole plant specific leaf area were also evident in the temporal dynamics of changes in this trait following the transfers between low and high light. In both species, leaf area ratio (LAR) declined in all treatments during the course of the experiment (Fig. 4a,c). (This decline probably reflects an ontogenetic trend, rather than an environmental effect, because light levels were also declining over this period, which would cause an increase rather than a decrease in LAR.) The responses of the plants following the experimental transfer can be illustrated by calculating the difference in LAR between each set of transferred plants and the control plants in the respective postswitch environment (i.e., HL versus LL

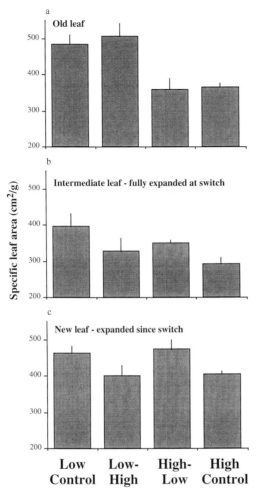

Figure 3 Specific leaf area (SLA) of individual leaves of *Heliocarpus appendiculatus* 10 days after reciprocal transfer between low- and high-light environments (N = 6, error bars = 1 s.e.). The SLA of mature leaves (a) did not change when leaves were transferred between low- and high-light environments. Leaves transferred from high to low (HL) were identical to high-light controls (HH), and LH leaves were identical to LL controls. Leaves that have just reached full expansion exhibit partial responses (b), and the SLA of the transferred leaves was intermediate between the two controls. New leaves expanded after plants were transferred (c) were identical to controls, exhibiting full acclimation. As a consequence, changes in SLA at the whole plant level depend in part on turnover of the leaf population, so the response time of this character is coupled with leaf life span.

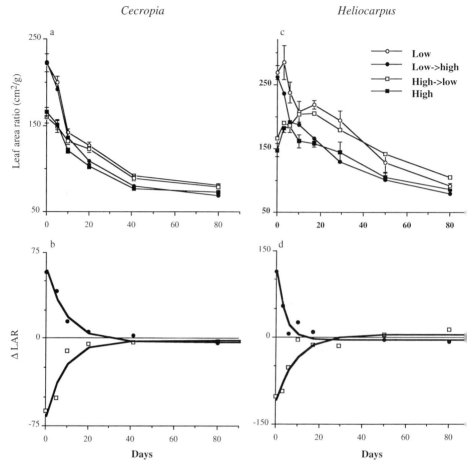

Figure 4   (a, c) Time course of leaf area ratio (LAR) following reciprocal transfers between low and high light, for *Cecropia obtusifolia* and *Heliocarpus appendiculatus*. Both species exhibited an overall ontogenetic decline in LAR during the 80 days following the switch in all four treatments. (b, d) Changes in LAR following the change in light environment are illustrated by calculating the difference between the transferred plants and the control plants in the postswitch environment; the upper line is LH − HH, and the lower one is HL − LL. These curves illustrate the asymptotic approach of switched plants to the LAR values expected in the new environment. A first-order kinetic equation was fit to these curves, along with similar ones for specific leaf area (see text); parameters of this equation are shown in Table II.

and LH versus HH). These differences confirm that the LAR of plants transferred between environments responded very rapidly and converged on the respective control treatments (Fig. 4b,d). Visual inspection suggested that these patterns could be described by a first-order kinetic equation of the form:

$$X_d = (X_0 - S)\ e^{-d/\tau} + S, \tag{1}$$

where X_d is the difference between the switch and control plants on day d, X_0 is the initial difference on the day of the switch, S is the asymptotic value approached in the postswitch environment, and τ is the time constant. The time constant reflects the time to exhibit approximately 65% of the final response. Values of X_0, S, and τ were determined from the six or eight harvest points by iterative nonlinear curve-fitting using the NONLIN module of SYSTAT (Wilkinson, 1987); for the four analyses (two species and two directions of change), this function explained 83 to 97% of variation in mean values at successive harvests. The time constants for SLA and LAR ranged from 3.0 to 9.6 days, with the lowest values being observed for *Heliocarpus* switched from low to high light, again emphasizing the rapidity of these responses (Table II).

As expected, proportional biomass allocation exhibited more persistent effects of the preswitch environment compared with leaf traits. For the support weight ratio [(stem + branch + petiole biomass)/total biomass], effects of the postswitch environment in *Cecropia* were not observed until 41 days after the transfer; in *Heliocarpus* postswitch effects were observed

Table II Parameters of First-Order Kinetic Functions
Fit to Responses of Specific Leaf Area and Leaf Area Ratio Following Switch[a]

Species	Parameter	Specific leaf area (cm²/g)		Leaf area ratio (cm²/g)	
		LH	HL	LH	HL
Cecropia	X_0	119	−131	59.8	−66.3
	S	18	−20	−2.57	−0.627
	τ	7.83	6.30	9.60	8.64
	r^2	0.95	0.97	0.97	0.91
Heliocarpus	X_0	209	194	113.5	−107.1
	S	−17	−1	−3.47	6.74
	τ	3.01	7.09	4.03	9.60
	r^2	0.83	0.90	0.93	0.87

[a] Values of both parameters were expressed as the difference between the switched treatment and the corresponding control, as illustrated in Fig. 4. The NONLIN module of SYSTAT was used to fit the following model:

$$X_d = (X_0 - S)\ e^{-d}/\tau + S$$

where X_d is the difference between the switch and control plants on day d, X_0 is the initial difference on the day of the switch, S is the asymptotic value approached in the postswitch environment, and τ is the time constant ($N = 48$ per treatment).

by day 6, but the effects of the preswitch environment persisted throughout the experiment up to 80 days following the transfer (Table I). Following the transfer to the new environments, the rate of height growth, measured in the intervals between harvests, responded to the new environment immediately following the switch (Table I). Total plant height, however, did not exhibit an effect of the postswitch environment until 40 or 50 days after the transfer, when the cumulative effects of changes in growth rate eventually outweighed the initial differences between plants in low and high light. The difference in the response time of height growth rate versus total plant height illustrates how a physiological or developmental process may respond immediately, while allocational and size-related characters that reflect the cumulative outcome of such processes will exhibit slower changes.

III. Plant Growth and Allocation in Fluctuating Light Environments

The rapid responses of leaf area display and to a lesser extent of biomass allocation suggest that oscillating environments on the scale of days to weeks could have significant effects on the growth of these species. For traits that respond to a change in environment within 3 to 10 days, fluctuating light levels in cycles more and less rapid than this time frame would be expected to have significant effects on allocation and leaf area display, and possibly also on total growth. In particular there is an intriguing possibility that intermediate scale oscillations will depress plant performance, because the adjustments in physiological and allocational traits following each change are out of phase with the environmental oscillations. An unusual experiment addressing photoperiodic responses provides potential evidence for such an effect at short time scales. Garner and Allard (1931) grew a variety of plants in continuously oscillating environments, with dark and light periods ranging from 5 s to 12 hr. In some species, total growth was reduced by an order of magnitude in oscillations of 1 to 15 min, relative to shorter *and* longer periods. This time scale is of the same order as the time constant of photosynthetic induction (Pearcy *et al.*, 1994), so the leaves of these plants may have oscillated between induced and noninduced states, always lagging behind the changes in light levels. Studies of plant growth in diurnally variable light environments have also observed higher growth under more rapid oscillations (Hogan, 1990; Sims and Pearcy, 1993), though growth is generally higher in steady-state environments with equivalent total light (Sims and Pearcy, 1993; Wayne and Bazzaz, 1993). On the basis of the response dynamics of specific leaf area observed in the experiment above, I first explored a simulation model to predict responses of this trait in slowly oscillating environments. The predictions of this simulation were then tested by growing *Heliocarpus* seedlings in fluctuating light environments with different periods and phases (defined below). The objectives of this experiment were (i) to assess the time scale at which different traits either track or integrate environmental fluctuations; (ii) to determine whether the characteristics of plants growing in rapidly fluctuating light environments resemble more closely those of plants growing in constant sun or shade, or are intermediate between the two; and (iii) to test whether the temporal scale and pattern of environmental heterogeneity influence plant performance independently of the total quantity of resource. Further details of experimental design and results are given in Ackerly (1993).

A. Predicting Responses to an Oscillating Environment

The dynamics of change in SLA following reciprocal transfer between light environments were used to model how this trait might behave in a fluctuating light environment, depending on the time scale of the fluctuations. Hypothetical plants were assigned an intermediate starting value of SLA, and then one was "placed" in low light and the other in high light. On the basis of experiments with *Heliocarpus* (which was subsequently used in the test of this model), low- and high-light values of SLA were set at 440 and 275 cm^2/g, respectively, and time constants of 3 and 7.1 days, for low to high and high to low transfers, respectively, were obtained from the results above. The simulated environment of each plant alternated between high and low light at a range of different frequencies, with periods ranging from 1 to 30 days in each environment, and the simulations covered a duration of 60 days. Additionally, contrasting phases were simulated at each frequency; one series of fluctuations started in high light and a paired one started in low light, such that on any particular day the two phases within a particular periodicity exhibited contrasting light environments. In the 30-day oscillations, the simulated plants attained the asymptotic values set for each of the two environments (Fig. 5a). In 10-day oscillations, however, the plants transferred from low to high light reach the expected asymptote, whereas those transferred from high to low light fall short of the asymptotic value because of the longer time constant for transfers in this direction (Fig. 5b). In rapid, 3-day fluctuations, the plants oscillate around an intermediate value, but this value is closer to the SLA expected in high light because the low- to high-light time constant is shorter (Fig. 5c). The predictions from this simulation can be summarized by plotting the values of SLA after 60 days in environments fluctuating at different periods, with one set of plants ending with a final period in low light and another ending in high light (Fig. 5d). For rapidly fluctuating environments the plants in the two phases exhibit similar final values, but they are closer to the asymptotic

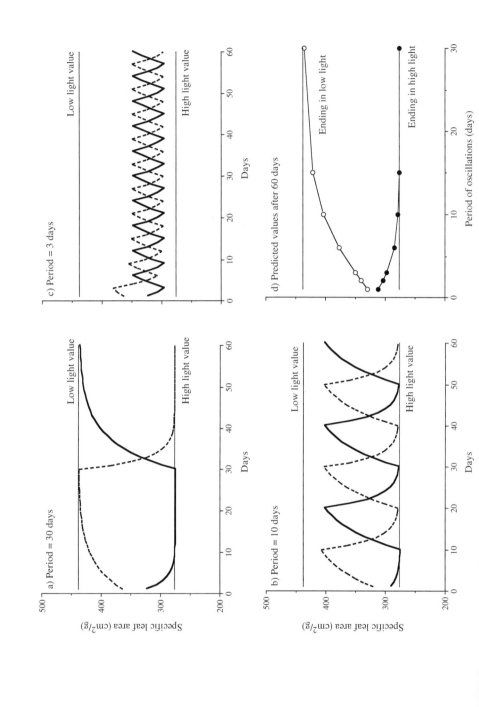

value for high light, because the trait change is more rapid following the low to high change. As the period lengthens, the plants that end in high light rapidly approach the asymptotic value, whereas those ending in low light reach the asymptote only in the longest period of 30 days.

B. Experiment II: Growth in Oscillating Light Environments

The second experiment presented here addresses the objectives listed above and provides a test of the predictions of this simulation. For this experiment, the treatments consisted of alternating periods of high and low light of different frequencies and phases; the two light environments averaged 3.7 and 10.5 mol m^{-2} day^{-1} over the course of the experiment, with daily values in high light ranging from 3 to 24 mol m^{-2} day^{-1} owing to fluctuations in weather conditions (Fig. 6a). There were seven different frequencies of temporal oscillations, with periods in low or high light of 1, 2, 3, 6, 10, 15, or 30 days, and a total experimental duration of 60 days. At each frequency there were two opposing phases, with one set of plants that started in low light and a paired set that started in high light; the seven frequencies and two phases created a total of 14 distinct treatments ($N =$ 6 per treatment). The plants were arranged in pairs, consisting of one from each phase from a particular frequency. Low-light conditions were created by constructing individual shade cloth covers that hung above the greenhouse bench. The alternations between light levels were achieved by moving the covers back and forth between the two plants in each pair, generating the opposing phase treatments while eliminating movement of the plants themselves. After a period of 60 days, each plant had experienced 30 days in the high-light environment and 30 days in low light, and the plants were harvested to determine final biomass, proportional allocation, and specific leaf area. Integrating light sensors were used to monitor light levels adjacent to every plant, documenting the alternating periods of low and high light (Fig. 6b–d). Because of a very fortunate symmetry in the pattern of shady and sunny days during the course of the experiment, the total, integrated photosynthetic flux density over the entire 60-day period was virtually identical across all 14 treatments, ranging from 400 to 450 mol m^{-2} (Fig. 7a).

Figure 5 (a–c) The dynamics of specific leaf area following transfers between high- and low-light environments were used to simulate responses of this character in a fluctuating environment, with different periods in sun and shade. The dynamics were generated using first-order kinetic responses with time constants determined in the reciprocal switch experiment. Dashed lines denote plants placed first in low light, and solid lines denote plants placed initially in high light. (d) The period of fluctuations influences the predicted final values after 60 days in fluctuating light. Owing to differences in the time constants of low–high and high–low transfers, the values expected in rapidly fluctuating environments are closer to the asymptotic value for high-light plants.

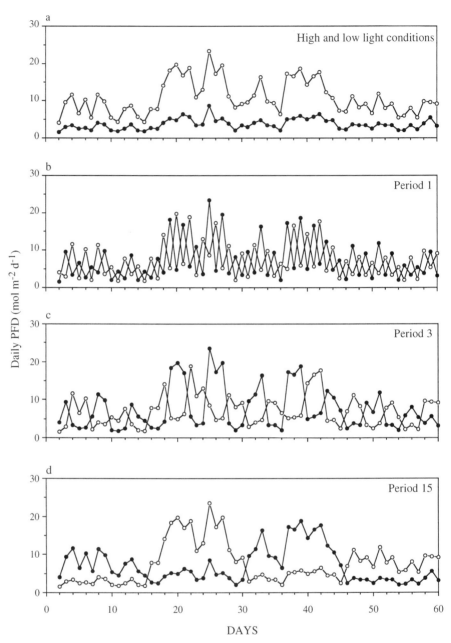

Figure 6 (a) Time course of daily light levels (PFD, photosynthetic flux density) in the high- and low-light conditions used for the oscillating light experiment. (b–d) Light environments in 6 of the 14 treatments in which plants were moved back and forth between low- and high-light conditions at seven different frequencies and in two different phases, with one group in low and another in high light on each day. The period refers to the number of days in each patch of low or high light.

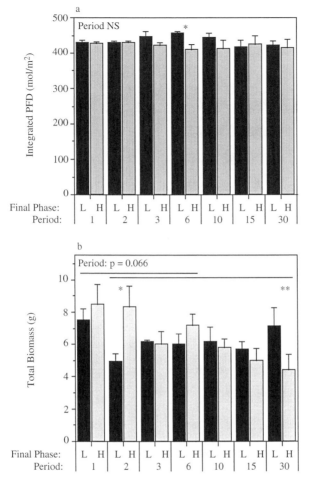

Figure 7   (a) Total photosynthetic flux density integrated over the 60-day period of the experiment was virtually identical in all 14 treatments. The seven frequencies are indicated by the respective periods, and the two phases are identified by the environment in which the plants were growing at the end of the experiment. (b) Final biomass was significantly greater in the more rapidly fluctuating environments, with periods of 1 day each in low and high light, than in slower fluctuations with periods of 10 days or more. The lines over the treatments connect frequencies that were not significantly different from one another. Stars indicate that the two phases were significantly different at a particular frequency: *, $p < 0.05$; **, $p < 0.01$; ***, $p < 0.001$.

Though there was no significant variation in total light received, there were significant differences in final plant biomass among the 14 treatments, with a trend toward lower biomass in fluctuations with longer periods

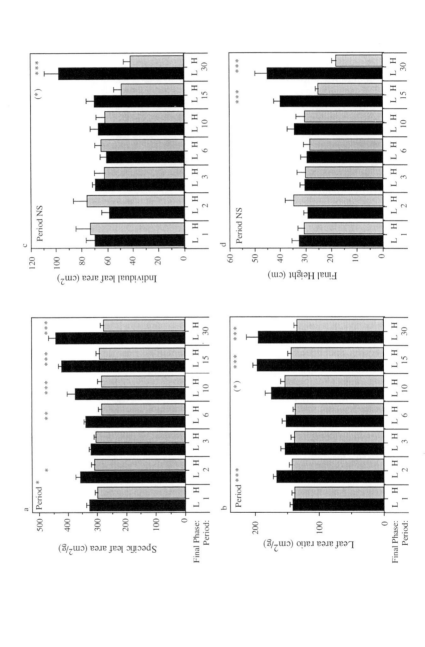

(Fig. 7b). Orthogonal contrasts indicated that total biomass in the 1 day treatment (the most rapid fluctuation) was significantly greater than that in the 10-, 15-, and 30-day periods. Additionally, there were significant differences between the plants in the opposing phases in the 2- and 30-day periods; in the latter case the plants that grew initially in high light and were then transferred to low light had a greater final biomass. These results clearly indicate that variation in the temporal scale and pattern of resource heterogeneity, on scales of a day and longer, can strongly influence whole plant carbon gain independently of total resource availability in the environment. The decline in growth at the longer periods parallels the results in diurnally variable environments in which less growth was observed under slower oscillations (see Garner and Allard, 1931; Hogan, 1990; Sims and Pearcy, 1993; Wayne and Bazzaz, 1993).

In addition to the differences in biomass, the period and phase (starting in low versus high light) of the fluctuations in light conditions had strong effects on biomass allocation, as well as leaf and canopy structure. These effects were assessed by a nested analysis of variance, with the phases nested within frequencies. As expected, strong phase effects were observed for the longest period, as these plants had been in their respective environments for 30 days at the time of harvest. The plants that ended in the shade were taller, with greater allocation to support biomass and lower allocation to roots, greater total leaf area, larger mean leaf size, higher specific leaf area and leaf area ratio, and higher leaf display efficiency (the ratio of projected canopy area to total plant leaf area, combining effects of leaf angle and overlap among leaves; Chazdon, 1985; Ackerly and Bazzaz, 1995; Fig. 8). As the frequency of environmental fluctuation increased, the differences between plants ending in sun and shade decreased or disappeared. For example, differences in mean individual leaf size were observed only at period 30, in final height at periods 15 and 30, in leaf area ratio at periods 10 through 30, and in specific leaf area at periods 2 and 6 through 30 (Fig. 8, Table III). At the highest frequency, with 1-day fluctuations between low

Figure 8 Specific leaf area (a), leaf area ratio (b), mean individual leaf area (c), and final height of plants (d) grown in oscillating light environments of different frequencies and phases (symbols as in Fig. 7). The differences between the plants in the two phases increased at slower frequencies, as the duration of time since the last transfer increased. Significant differences between the two phases indicate that the plants were tracking environmental fluctuations at a particular frequency. Traits that respond more rapidly following a change in the environment, such as specific leaf area, track fluctuations on shorter time scales. Mean individual leaf area, which changes more slowly since it depends on production of new leaves, responds more slowly and only tracks fluctuations on long time scales of 15 to 20 days (see Table III).

Table III   Summary of Phase Effects for Characters of Plants Growing in Fluctuating Light Environments of Different Frequencies, Indicated by Length of Each Period in Low or High Light[a]

Trait	Length of period (days)						
	1	2	3	6	10	15	30
Total biomass		*					**
Height						***	***
Leaf weight ratio					***		*
Support weight ratio						*	***
Root weight ratio							***
Total leaf area						*	***
Projected leaf area		**			*	***	***
Leaf display efficiency			*	*	*	*	**
Individual leaf area							***
Specific leaf area		*		**	***	***	***
Leaf area ratio						***	***
Number of traits	0	3	1	2	4	7	11

[a] A significant result indicates that the plants in the two phases, one in low and one in high light, exhibited significant differences at the final harvest, after 60 days of growth. Projected leaf area is the area of the seedling canopy when viewed from directly overhead, and leaf display efficiency is the ratio of projected leaf area to total plant leaf area. *, $p < 0.05$; **, $p < 0.01$; ***, $p < 0.001$.

and high light, the characteristics of plants ending in sun and shade were indistinguishable.

The observed values of specific leaf area correspond extremely well to the predictions based on the kinetics of response following the reciprocal transfer (Fig. 9). Note that the exact correspondence of the values at period 30 occurred because the asymptotic values of SLA used in the simulation were based on the values observed in the 30-day fluctuation in this experiment. For the other frequencies, the observed and predicted values correspond in two important regards: (i) the plants ending in low light diverge from the asymptotic values at slower frequencies than those ending in high light (e.g., at periods 6, 10, and 15), and (ii) the value of SLA in the 1-day fluctuations is not halfway between the equilibrium values but is closer to the asymptotic value observed in high light. In the simulation, these patterns emerge simply from the difference in the time constants for transfers in the two directions. The functional significance of the changes in SLA was not directly determined in this experiment. It is possible that the patterns reflect changes in carbon gain, as carbohydrates are accumulated and translocated at different rates in low and high light. Regardless of the mechanism, differences in the time constant for different directions of

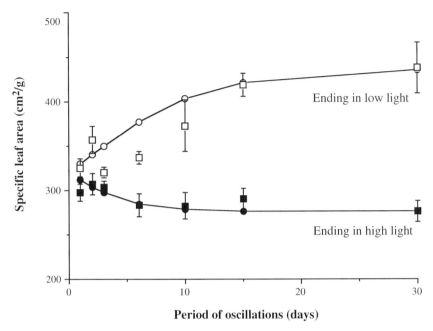

Period of oscillations (days)

Figure 9 Predicted and observed values of specific leaf area after 60 days of growth in fluctuating light environments were in close agreement in two important regards: (i) plants ending in high light reached asymptotic values in more rapid oscillations than those ending in low light, and (ii) in rapidly fluctuating environments the final value of SLA was closer to the asymptotic value in high light than in low light. The correspondance of the asymptotic values themselves was set in the model and not predicted *a priori*.

environmental change will result in asymmetric responses when environments fluctuate on time scales shorter than the time constants of response.

Functionally, leaf carbon gain may be enhanced in a rapidly fluctuating environment if the leaf takes on appropriate physiological characteristics for the patch that provides greater potential for photosynthesis, namely, the high-light environment (see Takenaka, 1989). Thus, the shorter time constant observed in switches from low to high light may result in rapid adjustment to periods of high energy availability, with slow return to the low-light phenotype. The time constant of response also provides an integrating mechanism allowing the plant to respond to average environmental conditions over particular time scales. In the simulation above, the response of SLA is determined each day in relation to the prevailing ambient conditions, either increasing or decreasing depending on its value relative to the equilibrium value in the current environment. And yet, if the environment is fluctuating rapidly, relative to the time constant of response for a particular trait, that trait will integrate or filter these fluctuations and exhibit a value intermediate between those observed in either steady-state condition.

IV. Tracking a Randomly Varying Environment

The role of the time constant as an integrating mechanism was explored further through simple simulations of a phenotypic trait tracking a randomly varying environment. The objective of this simulation was to generate a time series of data points for a hypothetical environmental condition, along with the responses of several traits that track the environment but that vary in the time constant of response. The value of the trait at each moment in time was then compared with the running mean of the environment over different periods of the recent past, in order to determine the time period over which the trait effectively integrates environmental heterogeneity. For convenience the values for the environment and the trait are measured on the same scale, and in each time step the trait value (P) changes, moving toward the environmental value (E), by the amount

$$\Delta P = \frac{E - P}{\tau} . \tag{2}$$

Thus, if the time constant (τ) is 1, the trait will reach its equilibrium value in each time step and track the environment perfectly. As the time constant is increased, the adjustments in the phenotype in each time step are progressively smaller, so the trait tracks environmental changes with a time lag and never reaches its equilibrium value.

The time sequence of environmental values was generated as a random walk, with the change in each time step selected at random within the interval ± 1. A trait with a time constant of 3 days tracks this random walk with no perceptible lag (Fig. 10a). In contrast, a trait with a time constant of 30 days exhibits a smoother fluctuation, lagging behind the current environment while reflecting conditions in the recent past (Fig. 10b). The correlation between the value of the trait at each moment in time and the running mean of past environmental values over different periods of time provides a measure of the interval over which the trait integrates environmental heterogeneity. For a trait with a time constant of 3, this correlation peaks, with an r^2 of almost 1, for running means with an interval of 4 days (Fig. 10c). In other words, the value of the trait at each moment in time reflects, with almost perfect correspondence, the mean of the past 4 days of environmental conditions. For traits with longer time constants, the peak correlation is observed at successively longer averaging intervals: 10 days for $\tau = 6$, 26 days for $\tau = 15$, and 48 days for $\tau = 30$. As above, these simulations illustrate that the response time of a trait acts as a mechanism for integrating environmental fluctuations, that is, as a filter that removes short-term oscillations and responds to fluctuations on the same time scale or longer as the time constant of response.

This approach, based on correlations of trait values with environmental running means, was applied to the results obtained in the oscillating light experiment. Rather than a time series of phenotypic values at different points in time, this experiment provides a set of phenotypic values that were obtained at one point in time (the final harvest), associated with a variety of prior patterns of environmental variation. For each plant, the mean value of the light environment was calculated for successively longer intervals of time prior to the harvest, from 1 to 59 days, and correlation coefficients were calculated between these means and the values of different traits at the harvest. This analysis demonstrates that the state of a given trait was correlated with the average conditions over some interval of time in the past, and the overall magnitude of correlations and the time interval with the greatest R value varied among traits (Fig. 11). Specific leaf area and leaf area ratio were both highly correlated with mean environmental conditions in the final 20 days before harvest ($R > 0.55$), with a maximum correlation at 15 days. Leaf display efficiency also exhibited fairly strong correlations with mean light environments averaged over 1 to 20 days, with two apparent peaks at 4 and 15 days. In contrast, root weight ratio was not correlated at all with light levels immediately prior to harvest, but exhibited a maximum correlation with conditions averaged over 20 days. The positions of the maximum correlations (Fig. 11) correspond approximately with the frequencies at which phase effects were detected for each trait (Table III). Leaf display efficiency exhibited a significant difference at periods of 3 days and longer, specific leaf area from 6 days and up, leaf area ratio at 15 and 30 days, and root weight ratio only at 30 days.

Many plant traits cannot be evaluated without considerable disturbance or destruction of the individual. These results suggest that it might be possible to assess the relative response times of multiple traits by combining continuous measurements of the environment for a number of plants with a single subsequent assessment of various phenotypic traits. In any natural situation, a difficult problem arises because traits are sensitive to multiple environmental factors, and their effects may not be additive. In simple cases, however, this sort of statistical approach may be able to distinguish the relative importance of different environmental factors on different traits, as well as the differences in response times among traits. Some traits may also have multiple time constants of response, reflecting several underlying processes, and further simulations would be necessary to determine whether analyses based on correlation coefficients would detect such processes.

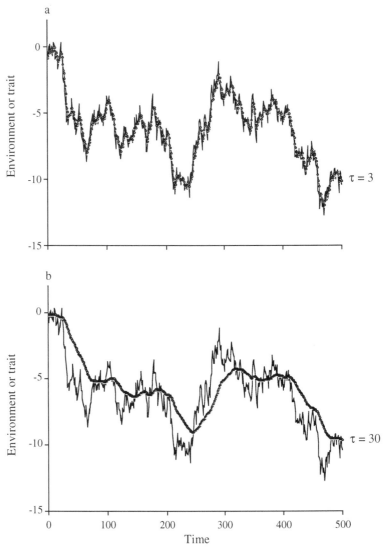

Figure 10   (a) Simulation of a randomly varying environment and the responses of a phenotypic trait with a time constant of 3 days. The trait tracks the environment very closely, and the two lines cannot be distinguished. (b) Similar simulation for a trait with a time constant of 30 days. The trait exhibits dampened responses and lags behind rapid changes in environmental conditions. (c) The correlation coefficient between running means of environmental conditions, averaging over bins of different duration, and traits with different time constants indicates the time scale on which the traits track the environment. A trait with a time constant of 3 days exhibits a maximum correlation with the environmental value averaged over the previous 4 days; for $\tau = 6$ the maximum correlation is with a 10-day interval, for $\tau = 15$ with a 26-day interval, and for $\tau = 30$ with a 48-day interval.

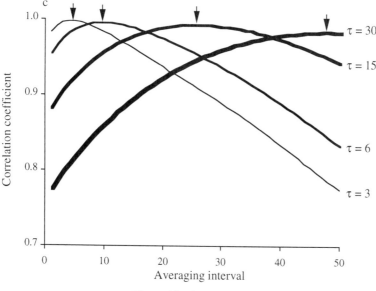

Figure 10 (*continued*).

V. Conclusions and Speculations

Under natural conditions, environmental heterogeneity occurs across a broad range of scales in space and time. For temporal variation, there are strong peaks in this variation at daily and annual time scales, as well as lesser peaks on shorter and longer scales, due to climatic patterns (Mitchell, 1976) and vegetation change. The effect on plants of variation at these different scales depends critically on the length of the life cycle, and the rate of response of physiological and ecological processes of interest. Year-to-year variation is experienced very differently by annual and perennial plants. For the former, the changes in the environment are encountered by successive generations, whereas for perennial plants each individual experiences the fluctuations and individual acclimation may be an important component of response and performance through time. For a particular species, similar considerations apply when examining the response of different traits. For example, specific leaf area and related traits measured at the whole plant level cannot track environmental changes from day to day (Fig. 8), whereas photosynthetic induction levels and related physiological traits can track these same fluctuations. These differences in response time reflect the various physiological and developmental mechanisms of plasticity in different traits. An understanding of these mechanisms sheds light on the relationships among responses of different traits and the varia-

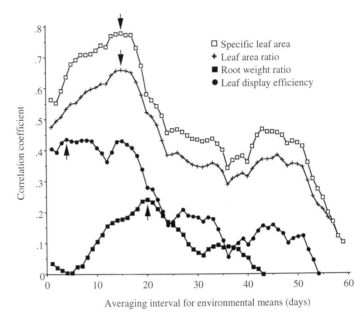

Figure 11   Using the data from the oscillating light experiment, correlation coefficients were calculated between the values of several traits, measured at the final harvest, and the mean light levels averaged over different intervals before the harvest. For leaf display efficiency, a trait with rapid response times, correlations were highest for 4- to 12-day averaging intervals; for specific leaf area and leaf area ratio maximum correlations were observed at 15 days, and for root weight ratio at 20 days, indicating that the latter traits respond more slowly to changing environmental conditions.

tion in response among species. For example, the potential rate of response of biomass allocation is strongly influenced by relative growth rate. This association may explain why, in switching experiments conducted over a particular time span, greater plasticity is observed in pioneer species than in slower growing, late successional species (e.g., Fetcher *et al.*, 1983; Strauss-Debenedetti and Bazzaz, 1991; cf. Bazzaz, 1987). Similarly, the response of leaf traits, when measured at the whole plant level, depends in part on leaf life span, so this parameter plays a central role in the response dynamics of this suite of characteristics.

The functional significance of physiological and allocational responses in varying environments is difficult to assess. To determine the significance of a particular response, experimentally, it is necessary to compare plants with greater and lesser responsiveness, which is difficult to achieve. Comparisons of species or populations that vary in the degree of response in any one character are confounded by the many other factors that differ between any two genetically distinct groups of plants. It has been possible to test the functional value of the entire suite of sun and shade responses by growing plants in low and high light and then assessing their performance in each of the environments through reciprocal switches. Sims and Pearcy (1994; Sims *et al.*, 1994) conducted a thorough study of this sort using *Alocasia macrorrhiza* and concluded that sun/shade acclimation enhances carbon gain per unit leaf area in the respective environments, supporting the notion that acclimation is functionally adaptive. However, carbon gain per unit weight in high-light environments is comparable for shade and sun acclimated plants (also see Rice and Bazzaz, 1989b). On the basis of empirical data and a detailed, mechanistic model, Sims and Pearcy attributed this result primarily to the decrease in leaf area ratio in high-light plants, which counteracts their greater photosynthetic capacity per unit area. Unfortunately, these tests are based on short-term analysis of growth rates following a change in environmental conditions, so this approach cannot be used to assess the functional role of the continuous adjustments that occur in a fluctuating environment. Another approach to the problem is the use of mechanistic models of physiology and carbon gain, which allow quantitative analyses of the role of individual traits. Such models have provided considerable insight into the contribution of individual trait responses to carbon gain in temporally complex sunfleck regimes (e.g., Gross *et al.*, 1991). Comparable models of whole plant growth (such as Sims *et al.*, 1994) may provide similar insight into the contribution of responses in different traits to performance in slowly oscillating environments, but these analyses have not been conducted.

In the meantime, it is tempting to speculate regarding the potential causes of the reduction in growth in longer periods observed here (Fig. 7b), though the pattern itself also deserves confirmation through additional experiments. Above I suggested that the growth reduction at periods of 1 to 15 min in Garner and Allard's (1931) experiment might have been caused by the decoupling of acclimation responses (in that case, photosynthetic induction) and light environment, owing to the lag time of the response. Similarly, if responses in leaf area ratio are the most important component of the sun/shade response, in terms of relative growth rates, then the 10- to 20-day response time of this trait may have caused the plants to be out of sync with environmental fluctuations on a similar or slightly longer time scale, reducing growth in the 10- to 30-day periods. One corollary of this hypothesis is that if the experiment were carried out over a longer time period, and additional frequencies could be tested, total biomass would increase again at long periods, where the adjustments in allocation would catch up with the environment and enhance carbon gain in each set of conditions before light levels changed again.

An additional implication of this argument is that selection might operate on response time itself. If a trait, such as photosynthetic induction or leaf area ratio, plays an important function in carbon gain in variable light environments, then it would be important for the response time of the trait to be either considerably slower or faster than dominant periods in the environmental signal. For example, if photosynthetic induction times were on the order of 6 to 12 hr, leaves would reach full induction just as light levels declined in the afternoon, and the plant would gain little from the enhancement of photosynthetic rates that accompany full induction. On the other hand, induction times of several days would allow plants to stay in a highly induced state over longer time periods. Such a plant would not incur a "desynchrony" cost, though the night respiration cost of maintaining induction would presumably make this strategy inefficient. As it is, induction times under 1 hr allow the plant to track diurnal fluctuations, or increased light levels in large sunflecks and gaps, while reducing respiratory costs in intervening low-light periods. The converse of this argument is that if a trait is under some constraint that prevents rapid responses, there may be selection against a large magnitude of plasticity in response to particular environmental conditions. Specifically, if an appropriate response could only be elicited very slowly, then it might be better for the plant not to respond at all to an environmental fluctuation that might not persist. This would be particularly true in response to a transient period of low light, where the maximum enhancement of carbon gain that would accompany acclimation is greatly limited by the low energy levels. Related arguments regarding the role of "mixed strategies" in spatially fine-grained environments have been presented by Lloyd (1984).

Returning to the questions posed in the introduction, the experiments presented here along with prior studies clearly show the association between the physiological and developmental mechanisms of phenotypic responses, on the one hand, and the time scale of these responses in temporally varying environments, on the other. Consideration of these mechanisms is important as they may generate correlations among different aspects of growth and plasticity, as well as constraints on the responses of particular traits, that influence the role these traits play in variable environments. As expected on *a priori* grounds, the relative scale of plant responses and environmental fluctuations determines whether a trait will "track" or "integrate" environmental fluctuations. At various time scales, the frequency of fluctuating light environments is observed to influence growth, independently of the total amount of light received. It is possible that this effect reflects a decoupling of trait acclimation and environment, owing to lag times in trait responses, but this hypothesis requires further exploration. Overall, I hope that this chapter reinforces the dynamic aspect of phenotypic plasticity throughout the life cycle of the individual plant, not only in the ontogeny of the response to a particular environment (Pigliucci and Schlichting, 1995), but also in the potential for phenotypic response to environmental changes. Some traits, such as seed germination or the initiation of flowering in a determinate annual, are irreversible and expressed only once in the life cycle of the individual. For these traits, plasticity must be expressed and also evolve at the population level, as related individuals encounter different environments in space or time. For many other traits, changes may be rapid and reversible, allowing continuous adjustment to environmental change. If so, plasticity may be expressed within the life cycle of an individual growing in a fine-grained environment. These responses may make important contributions to the maintenance of plant performance and may also play an important role in the evolution of plasticity as an adaptive trait, and the relative importance of fine-grained and coarse-grained environmental variation in the evolution of plasticity deserves further study (see Lloyd, 1984; Gomulkiewicz and Kirkpatrick, 1992; Moran, 1992).

VI. Appendix

A simple quantitative analysis demonstrates that the response time of biomass allocation is closely coupled with plant relative growth rates. Consider a plant of total biomass B_t and root biomass B_r; proportional allocation of standing biomass to roots (root weight ratio) is $f_r = B_r/B_t$ (symbols follow Reynolds and Thornley, 1982). The plant is growing at a relative growth rate of μ, and a fraction λ of the new biomass gained at each point in time is partitioned into new root growth. The question we want to answer is, What is the rate of change in root weight ratio as a function of growth rate? The growth rates of total and root biomass are:

$$\frac{dB_t}{dt} = \mu B_t, \tag{A1}$$

$$\frac{dB_r}{dt} = \mu \lambda B_r. \tag{A2}$$

The rate of change in root fraction, f_r, is thus

$$\frac{df_r}{dt} = \frac{d(B_r/B_t)}{dt}. \tag{A3}$$

Applying the chain rule, this simplifies to

$$\frac{df_r}{dt} = \mu(\lambda - f_r). \tag{A4}$$

Therefore, the rate of change in standing allocation, under this model, is directly proportional to plant relative growth rate, and to the difference between the partitioning coefficient and the current allocation fraction.

Acknowledgments

I thank the editors for the invitation to contribute this chapter and Fakhri A. Bazzaz for support throughout this work. I am grateful to Sonia Sultan for insightful comments on the manuscript and to Louis Gross, Doug Karpa, Silvia Strauss-Debenedetti, Suzanne Morse, Erik Veneklaas, and Peter Wayne for discussions and comments that contributed greatly to this work, though none of them bear responsibility for faults that remain. P. Cannata, M. Hexner, C. DiMatteo, C. Smith, D. Smith, and B. Traw provided excellent assistance in the greenhouse. This research was supported by grants from the U.S. Department of Energy and the National Science Foundation.

References

Ackerly, D. D. (1993). Phenotypic plasticity and the scale of environmental heterogeneity: Studies of tropical pioneer trees in variable light environments. Ph.D. thesis, Harvard University, Cambridge, Massachusetts.

Ackerly, D. D. (1996). Canopy structure and dynamics: Integration of growth processes in tropical pioneer trees. *In* "Tropical Forest Plant Ecophysiology" (S. S. Mulkey, R. L. Chazdon, and A. P. Smith, eds.), pp. 619–658. Chapman & Hall, London.

Ackerly, D. D., and Bazzaz, F. A. (1995). Seedling crown orientation and interception of diffuse radiation in tropical forest gaps. *Ecology* **76**, 1134–1146.

Allen, T. F. H., and Starr, T. B. (1982). "Hierarchy: Perspectives for Ecological Complexity." Univ. of Chicago Press, Chicago.

Bazzaz, F. A. (1987). Experimental studies on the evolution of niche in successional plant populations. *In* "Colonization, Succession and Stability" (A. J. Gray, M. J. Crawley, and P. J. Edwards, eds.), pp. 245–272. Blackwell, Oxford.

Bazzaz, F. A., and Carlson, R. W. (1982). Photosynthetic acclimation to variability in the light environment of early and late successional plants. *Oecologia* **54**, 313–316.

Bongers, F., Popma, F., and Iriarte-Vivar, S. (1988). Response of *Cordia megalantha* Blake seedlings to gap environments in tropical rain forest. *Funct. Ecol.* **2**, 379–390.

Bradshaw, A. D. (1965). Evolutionary significance of phenotypic plasticity in plants. *Adv. Genet.* **13**, 115–155.

Caldwell, M. M., and Pearcy, R. W., eds. (1994). "Exploitation of Environmental Heterogeneity by Plants: Ecophysiological Processes Above- and Belowground." Academic Press, San Diego.

Canham, C. D. (1985). Suppression and release during canopy recruitment in *Acer saccharum*. *Bull. Torrey Bot. Club* **112**, 134–145.

Chazdon, R. L. (1985). Leaf display, canopy structure, and light interception of two understory palm species. *Am. J. Bot.* **72**, 1493–1502.

Chazdon, R. L. (1988). Sunflecks and their importance to forest understory plants. *Adv. Ecol. Res.* **18**, 1–63.

Evans, G. C., and Hughes, A. P. (1961). Plant growth and the aerial environment. I. Effect of artificial shading on *Impatiens parviflora*. *New Phytol.* **60**, 150–180.

Fetcher, N., Oberbauer, S. F., and Strain, B. R. (1983). Effects of light regime on the growth, leaf morphology, and water relations of seedlings of two species of tropical trees. *Oecologia* **58**, 314–319.

Fetcher, N., Oberbauer, S. F., Rojas, G., and Strain, B. R. (1987). Efectos del régimen de luz sobre la fotosíntesis y el crecimiento en plántulas de árboles de un bosque lluvioso tropical de Costa Rica. *Rev. Biol. Trop.* **35**(Suppl. 1), 97–110.

Garner, W. W., and Allard, H. A. (1931). Effect of abnormally long and short alternations of light and darkness on growth and development of plants. *J. Agric. Res.* **42**, 629–651.

Gomulkiewicz, R., and Kirkpatrick, M. (1992). Quantitative genetics and the evolution of reaction norms. *Evolution* **46**, 390–411.

Grime, J. P., and Campbell, B. D. (1991). Growth rate, habitat productivity, and plant strategy as predictors of stress response. *In* "Response of Plants to Multiple Stresses" (H. A. Mooney, W. E. Winner, and E. J. Pell, eds.), pp. 143–159. Academic Press, San Diego.

Gross, L. J. (1986). Photosynthetic dynamics and plant adaptation to environmental variability. *Lect. Math. Life Sci.* **18**, 135–170.

Gross, L. J., Kirschbaum, M. U. F., and Pearcy, R. W. (1991). A dynamic model of photosynthesis in varying light environments taking account of stomatal conductance, C3-cycle intermediates, photorespiration and RuBisCo activation. *Plant Cell Environ* **14**, 881–893.

Hogan, K. P. (1990). Acclimation of photosynthetic induction responses to different sunfleck regimes, and the consequences for plant growth. *Bull. Ecol. Soc. Am.* **71**(Suppl.), 191.

Jurik, T. W., Chabot, J. F., and Chabot, B. F. (1979). Ontogeny of photosynthetic performance in *Fragaria virginiana* under changing light regimes. *Plant Physiol.* **63**, 542–547.

Koike, T., Sanada, M., Lei, T. T., Kitao, M., and Lechowicz, M. J. (1992). Senescence and the photosynthetic performance of individual leaves of deciduous broadleaved trees as related to forest dynamics. *In* "Research in Photosynthesis, Volume IV" (N. Murata, ed.), pp. 703–706. Kluwer Academic Publishers, Dordrecht, The Netherlands.

Kuiper, D., and Kuiper, P. J. C. (1988). Phenotypic plasticity in a physiological perspective. *Acta Oecol. Oecol. Plant.* **9**, 43–59.

Levins, R. (1968). "Evolution in Changing Environments." Princeton Univ. Press, Princeton, New Jersey.

Lloyd, D. G. (1984). Variation strategies of plants in heterogeneous environments. *Biol. J. Linn. Soc.* **21**, 357–385.

MacArthur, R., and Levins, R. (1964). Competition, habitat selection, and character displacement in a patchy environment. *Proc. Natl. Acad. Sci. U.S.A.* **51**, 1207–1210.

Mitchell, J. M. J. (1976). An overview of climatic variability and its causal mechanisms. *Quat. Res. (N.Y.)* **6**, 481–493.

Moran, N. (1992). The evolutionary maintenance of alternative phenotypes. *Am. Nat.* **139**, 971–989.

Oberbauer, S., and Strain, B. (1985). Effects of light regime on the growth and physiology of *Pentaclethra macroloba* (Mimosaceae) in Costa Rica. *J. Trop. Ecol.* **1**, 303–320.

Osmond, C., Oja, V., and Laisk, A. (1988). Regulation of carboxylation and photosynthetic oscillations during sun–shade acclimation in *Helianthus annuus* measured with a rapid-response gas exchange system. *Aust. J. Plant Physiol.* **15**, 239–51.

Osunkoya, O. O., and Ash, J. E. (1991). Acclimation to a change in light regime in seedlings of six Australian rainforest tree species. *Aust. J. Bot.* **39**, 591–605.

Pearcy, R. W., and Sims, D. A. (1994). Photosynthetic acclimation to changing light environments: Scaling from the leaf to the whole plant. *In* "Exploitation of Environmental Heterogeneity by Plants" (M. M. Caldwell and R. W. Pearcy, eds.), pp. 145–174. Academic Press, San Diego.

Pearcy, R. W., Chazdon, R. L., Gross, L. J., and Mott, K. A. (1994). Photosynthetic utilization of sunflecks: A temporally patchy resource on a time scale of seconds to minutes. *In*

"Exploitation of Environmental Heterogeneity by Plants" (M. M. Caldwell and R. W. Pearcy, eds.), pp. 175–208. Academic Press, San Diego.

Pigliucci, M., and Schlichting, C. D. (1995). Ontogenetic reaction norms in *Lobelia siphilitica* (Lobeliaceae): Response to shading. *Ecology* **76**, 2134–2144.

Popma, J., and Bongers, F. (1991). Acclimation of seedlings of three tropical rain forest species to a change in light availability. *J. Trop. Ecol.* **7**, 85–97.

Reich, P. B., Walters, M. B., and Ellsworth, D. S. (1992). Leaf life-span in relation to leaf, plant, and stand characteristics among diverse ecosystems. *Ecol. Monogr.* **62**, 365–392.

Reynolds, J. F., and Thornley, J. H. M. (1982). A shoot:root partitioning model. *Ann. Bot.* **49**, 585–597.

Rice, S., and Bazzaz, F. A. (1989a). Quantification of plasticity of plant traits in response to light intensity: Comparing phenotypes at a common weight. *Oecologia* **78**, 502–507.

Rice, S., and Bazzaz, F. A. (1989b). Growth consequences of plasticity of plant traits in response to light conditions. *Oecologia* **78**, 508–512.

Robertson, G. P., and Gross, K. L. (1994). Assessing the heterogeneity of belowground resources: Quantifying pattern and scale. *In* "Exploitation of Environmental Heterogeneity by Plants" (M. M. Caldwell and R. W. Pearcy, eds.), pp. 237–253. Academic Press, San Diego.

Shultz, L. M., Rundel, P. W., Mulkey, S. K., and Wright, S. J. (1989). Leaf anatomy and carbon isotopes in a tropical rain forest: Watering response of an understory shrub. *Association of Tropical Biology Abstracts*, 30.

Sims, D. A., and Pearcy, R. W. (1991). Photosynthesis and respiration in *Alocasia macrorrhiza* following transfers to high and low light. *Oecologia* **86**, 447–453.

Sims, D. A., and Pearcy, R. W. (1992). Response of leaf anatomy and photosynthetic capacity in *Alocasia macrorrhiza* (Araceae) to a transfer from low to high light. *Am. J. Bot.* **79**, 449–455.

Sims, D. A., and Pearcy, R. W. (1993). Sunfleck frequency and duration affects growth rate of the understorey plant, *Alocasia macrorrhiza*. *Funct. Ecol.* **7**, 683–689.

Sims, D. A., and Pearcy, R. W. (1994). Scaling sun and shade photosynthetic acclimation in *Alocasia macrorrhiza* to whole-plant performance—I. Carbon balance and allocation at different daily photon flux densities. *Plant Cell Environ.* **17**, 881–887.

Sims, D. A., Gebauer, R. L. E., and Pearcy, R. W. (1994). Scaling sun and shade photosynthetic acclimation of *Alocasia macrorrhiza* to whole-plant performance—II. Simulation of carbon balance and growth at different photon flux densities. *Plant Cell Environ.* **17**, 889–900.

Strauss-Debenedetti, S. I., and Bazzaz, F. A. (1991). Plasticity and acclimation to light in tropical Moraceae of different successional positions. *Oecologia* **87**, 377–387.

Takenaka, A. (1989). Optimal leaf photosynthetic capacity in terms of utilizing a natural light environment. *J. Theor. Biol.* **139**, 517–529.

Thornley, J. H. M. (1991). A model of leaf tissue growth, acclimation and senescence. *Ann. Bot.* **67**, 219–228.

Wayne, P., and Bazzaz, F. A. (1993). Birch seedling responses to daily time courses of light in experimental forest gaps and shadehouses. *Ecology* **74**, 1500–1515.

Wilkinson, L. (1987). "SYSTAT: The System for Statistics." SYSTAT, Inc., Evanston, Illinois.

11

Allocation Theory and Chemical Defense

Manuel Lerdau and Jonathon Gershenzon

Theory and practice in the calculation of the costs of plant chemical defenses have remained largely divorced, although both have seen important advances in the last few years. Ecological models of allocation now explicitly consider evolutionary forces, and even very physiological/mechanistic models are constructed within an evolutionary framework. This common framework has allowed the development, comparison, and use of these models in evaluating empirical studies of the costs of allocation. Concurrent advances in plant biochemical and structural analyses make it possible to calculate the full cost of allocation to chemical defenses. Ecologists can now interpret large-scale patterns of variation in the allocation to chemical defenses in the light of biochemical costs and evolutionary forces.

I. Life History and Ecophysiological Models

Studies on the allocation of resources by plants to chemical defenses have demonstrated that simple models based on principles of evolutionary biology can be used to describe the patterns of allocation to defense shown by many plants (Rhoades, 1979; Mckey, 1979; Coley *et al.*, 1985; Herms and Mattson, 1992; Zangerl and Bazzaz, 1992). These studies on allocation to defense have shed light on general theories of resource allocation by plants and have helped bring together previously disparate aspects of plant ecology. Recent advances in the calculations of the costs of chemical defenses have demonstrated the importance of ecological phenomena not always considered in cost/benefit analyses and have underscored the need to consider the full range of costs in the production and maintenance of defensive compounds.

Theories of plant chemical defense made major steps forward with the recognition by Feeny (1976) that the risk of herbivore damage should be incorporated into models of defense and with the application of optimality principles, originally applied in ecology to questions of foraging theory, to help explain patterns of chemical defense seen in plants (Rhoades and Cates, 1976; McKey, 1979). More recent theories of plant chemical defense have been based on integrating Feeny's apparency theory into some notion of the risk that plants face in terms of herbivory, the costs of allocation to defense in terms foregone growth, and the benefits in terms of tissue protected (Herms and Mattson, 1992; Zangerl and Bazzaz, 1992; Lerdau *et al.*, 1994a).

Plants allocate resources for three primary reasons: (1) for reproduction, to flowers, fruits, and seeds; (2) for resource capture and the production of new tissues, roots, leaves, and stems (to ensure that leaves can capture light); and (3) for the protection of already captured resources, that is, mechanical and chemical defenses. A cost/benefit approach to the analysis of allocation is tempting because allocation to one of the above three appears to imply reduced allocation to the others. Such an analysis can be performed at many different time scales and levels of biological organization because the impacts of allocation decisions occur across the entire biology of the organism.

Cost/benefit analyses are based on the physical law, "There is no free lunch." They treat axiomatically the notion that any increase in allocation to one possible sink leads to a decrease in allocation to other sinks. This principle has received its strongest ecological support in studies of animal foraging (Krebs and McCleery, 1984). Plant ecologists have only come on the notion more gradually, although Cohen (1966) introduced it years ago to resource allocation theory. One major caveat must be considered when using cost/benefit analysis: the trade-off among resource sinks is an assumption that, if it is not met, negates the entire analysis. This assumption requires empirical justification each time it is made.

Theories on the allocation of resources to defense have developed along two parallel but, until recently, largely separate lines. From the more obviously evolutionary perspective of life history strategy have come optimality models of allocation to defense in which the fitness of the plant is considered a function of the relative allocation of resources to defense versus reproduction. These models stress that the optimal allocation is that which maximizes the plant's lifetime reproductive value. Taylor et al. (1974) defined a trade-off between reproduction and survival:

$$\text{Max } V_{\text{fitness}} = \int e^{-rt} \, l_t b_t \, dt,$$

where r is the per capita rate of increase, l_t is the survival until age t, b_t is fecundity at age t, and t and x are age. However, this theory requires that reproduction at each time t be measured, a requirement that is not easily met for many plants (see below).

From the realm of environmental physiology has come a more mechanistic set of models, but ones still based on optimality principles, in which biomass is the quantity maximized with the trade-offs being between allocation to resource-gathering versus resource-protecting (defense) tissues. The three main ecophysiological theories have been recently reviewed (Gershenzon, 1994a) and are summarized below. The most explicitly evolutionary of these theories is the resource availability hypothesis of Coley *et al.* (1985). The hypothesis attempts to explain patterns of interspecific differences in allocation to defense and growth. Species native to resource-poor environments will have low growth rates and will tend to have high concentrations of immobile defenses such as condensed tannins; in contrast, species found in resource-rich habitats will show high growth rates and will utilize defensive compounds that can be metabolized easily such as alkaloids.

The carbon–nutrient balance hypothesis (CNBH; Bryant *et al.*, 1983) and the growth–differentiation balance hypothesis (GDBH; Loomis, 1932, as cited in Lorio, 1986) treat plants on ecological and physiological time scales, that is, discussing how individual plants respond to variations in their environment rather than how plants adapted to different environments vary in their allocation patterns. These two theories address questions of intraspecific variability in allocation to defense. The CNBH predicts that when a resource such as nitrogen is rare, then a plant will allocate more abundant resources, such as carbon, both to the acquisition (root growth) and protection (defensive compound production) of the rare resource. The GDBH makes similar predictions about the effects of resource scarcity and abundance, but it bases these predictions on the physiological assumptions that growth is more sensitive than photosynthesis to the availability of nutrients and that growth will be favored whenever resources are available. The GDBH predicts that when a belowground resource such as nitrogen is moderately limiting, then growth will be more constrained than carbon assimilation, and the photosynthate accumulated in excess of what can be used for growth will go toward defense.

Perrin (1993) and Lerdau (1993) have discussed the mathematical similarity and some of the operational differences between the life history strategy and ecophysiological approaches. The most important similarity lies in both approaches' being explicitly evolutionary in their underpinnings. Both rely on a link between allocation patterns and fitness. The difference lies in the currency that each uses to relate allocation patterns to fitness. The models based on life-history strategy tend to use fitness as their currency or to use seed production as a fitness surrogate. The physiological models tend to operate one step further removed by using biomass or the mass quantity of some particular element as a fitness surrogate. This difference in currency leads to important differences in the meaning of any experimental tests of the models. Different modeling approaches will be better suited for different biological questions.

The fundamental assumption of the life history models, that fitness is proportional to seed production, and the hypothesis examined by these models, that seed production is inversely proportional to defense, makes annual plants, where reproductive output in a year is probably a good correlate to fitness, ideal organisms for study. However, these two relationships make life history models difficult to use on other types of plants, especially long-lived woody perennials whose reproductive output over even 5 years may bear little relationship to the fitness of that plant. This problem is analogous to the question of reproductive output in iteroparous animals (Stearns, 1989). For a long-lived iteroparous animal, reproductive output in any one year may tell very little about the lifetime success of that animal. For many plants this problem is even more severe because they may live for a very long time, sometimes hundreds or thousands of years, as reproductively mature individuals. For all plants except annuals and those few short-lived perennials that are amenable to experimental manipulation and can be studied for their entire reproductive lifetimes, there is probably no way to proceed on allocation studies but to use ecological/physiological theories that use the fundamental assumption of fitness' being proportional to biomass and examine the hypothesis that seed production is inversely proportional to defense.

Both modeling strategies use trade-offs and future discounts as the conceptual bases for their development and mathematical structure. Both are based on the notion that allocation to one *telos* (reproduction, resource capture, or resource defense) leads to reduced allocation to the other two. Their use of future discounts stems from the recognition that a benefit in the present is "worth more" than the same quantity of benefit in the future. In ecophysiological terms, resource captured now allows growth and the capture of even more resources in the future. From the life history perspective, early reproduction is worth more because one's offspring can then breed earlier. The trade-off comes in the cost of not allocating to defense and thus increasing susceptibility to herbivores and/or pathogens (ecophysiological perspective) or reducing survivorship, l_t (life history viewpoint).

The mathematics for the calculations of optimal allocation are identical to those of standard economic models of investment under compounding interest. However, there is one crucial difference. Interest rates in economic models are derived from the government fixed rate for risk-free interest, that which is given by federal bonds (see, e.g., Rothschild and Stiglitz, 1970). Investments with increased interest rates will have increased risk. Because the discount rate equals the interest rate, situations with extremely high risk will have high discounting and proportionately more allocation to biomass (or reproduction) rather than defense (Lerdau, 1992). Nature, however, provides no risk-free environments; thus it is difficult, perhaps impossible, to compare observed discount rates to some theoretical optimum. This fundamental difference between ecological/evolutionary and economic models makes it difficult to apply the economic approach to biological systems because there is no risk-free interest rate to serve as a baseline for comparison. Despite this difficulty, relative comparisons among strategies can be made, but interpretations of biological meaning from economic models must be cautious.

II. Total Cost Calculations

All cost/benefit analyses, no matter how they deal with questions of discounting, require estimations of the total cost of producing and maintaining a defensive compound. For the life history-based theories one need only look at seed production differences between plants with different levels of defense (Simms and Rausher, 1987). However, as stated above, such an approach is impractical for all plants except annuals. For perennial plants, we need to estimate the cost of defense using a material currency such as carbon, nitrogen, or energy units. The exact currency used is not important as long the exchange ratios among currencies can be calculated (Koide and Elliot, 1989).

The full cost of a defensive compound requires consideration of the costs of five factors: (1) biosynthesis, (2) storage, (3) transport, (4) maintenance, and (5) foregone resource capture. Unfortunately, we know little about the biochemistry and physiology of most types of plant defense compounds, making accurate assessment their costs extremely difficult. A prominent exception is the monoterpenes. Advances in our understanding of metabolism, sequestration, and turnover provide sufficient information to permit realistic estimates of their costs.

Monoterpenes are 10-carbon members of the terpenoid family of natural products. They are most widely recognized as constituents of the Pinaceae, Myrtaceae, Rutaceae, Lamiaceae, and Asteraceae, although they have been recorded from nearly 50 families of higher plants (Banthorpe and Charlwood, 1980). Many monoterpenes have been demonstrated to play important roles as plant defenses, acting as toxins, feeding deterrents, and oviposition inhibitors to herbivores and as inhibitors of pathogen growth

(Gershenzon and Croteau, 1993; Harborne and Tomas-Barberan, 1991; Langenheim, 1994). In this section we describe the calculation of the cost of monoterpenes in the foliage of pines (*Pinus* spp.).

The biosynthetic cost of any plant metabolite has two components: (a) the cost of the required substrates and cofactors and (b) costs of the enzymes and other macromolecular machinery. Substrate and cofactor costs can be calculated directly from knowledge of the biosynthetic pathways using procedures developed by Penning de Vries, Mooney, and co-workers (Merino *et al.*, 1984; Mooney, 1972; Penning de Vries *et al.*, 1974). In these methods, cost is determined by computing the amount of glucose needed to provide all of the substrates (such as acetyl-CoA) and cofactors (such as ATP and NADPH) consumed in the biosynthesis. Figure 1 outlines the calculation of the substrate and cofactor costs for the formation of monoterpenes in *Pinus*. For each of these compounds, the synthesis of 1 g monoterpene requires 3.54 g of glucose. Because monoterpene biosynthesis in pines is restricted to a single specialized cell type (Bernard-Dagan *et al.*, 1979; Fahn and Benayoun, 1976), the costs of the enzymes and other machinery are not calculated directly, but are instead treated below as part of the cost of entire cells. It is difficult to determine independently monoterpene enzyme costs because of our lack of knowledge of enzyme turnover rates (Gershenzon, 1994a).

Monoterpenes in pine foliage are synthesized and stored in association with complex secretory structures called resin ducts (Fahn, 1979; Werker and Fahn, 1969). Each needle possesses a series of resin ducts found in the outer cortex that are parallel to the long axis of the needle. Monoterpenes are synthesized in the epithelial cells that surround the cortex and then released directly into the duct lumen. The construction and maintenance of the resin ducts and epithelial cells require a substantial commitment of resources. Both the ducts and epithelial cells contain thick, often suberized, walls, and the epithelial cells are highly differentiated and lacking in photosynthetic machinery and so require a constant supply of fixed carbon for growth and physiological function.

The costs of monoterpene storage in pines were deduced (Fig. 2) from a previously published estimate for the construction cost of pine foliage (Chung and Barnes, 1977) by computing the percentage of foliage represented by the epithelial cells and walls of the resin ducts as determined from cross-sectional micrographs of pine needles (Fahn, 1979). The costs for both epithelial cells and resin duct wall construction were assigned to the manufacture of resin constituents, because these cell types and walls are highly specialized for resin synthesis and storage. About half of these costs can then be charged to monoterpenes, since monoterpenes make up approximately 50% of the total resin constituents (Croteau and Johnson, 1985).

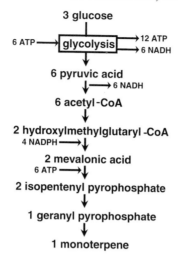

For each mole of monoterpene:

	moles required in pathway	moles made in pathway	net moles required	cost per mole in glucose units	net cost in glucose units
glucose	3	0	3	1	3
NADPH	4	0	4	0.086	0.34
NADH	0	12	(-12)	0.056	(-0.67)
ATP	12	12	0	0.028	0

total   2.67

$$\frac{2.67 \text{ moles glucose}}{\text{mole monoterpene}} \times \frac{\text{mole monoterpene}}{136 \text{ g}} \times \frac{180 \text{ g}}{1 \text{ mole glucose}} = \frac{3.54 \text{ g glucose}}{\text{g monoterpene}}$$

Figure 1 Pathway for biosynthesis of monoterpenes and calculation of costs. ATP is assumed to come from glycolysis and the TCA cycle coupled to mitochondrial electron transport. Since 1 mol glucose produces 36 mol ATP, the cost of one ATP is $1/36 = 0.028$ glucose unit. NADPH is assumed to arise from the pentose phosphate pathway, which makes 12 mol NADPH from each molecule of glucose with an investment of one ATP. Hence, the cost of one NADPH is $(1 + 1/36)/12 = 0.086$ glucose unit. NADH is manufactured in the course of glycolysis, so it is necessary to subtract its value from the total cost of terpene synthesis deduced. Since one NADH yields two ATP after transport to the mitochondrion, the cost of one NADH is $2 \times 0.028 = 0.056$ glucose unit. For more details, see Gershenzon (1994b).

Three components of the cost of defensive compounds, namely, transport, maintenance, and foregone photosynthesis, make only a small contribution to the total cost of monoterpenes in pines. Transport costs are

Cost of construction of pine foliage (Chung and Barnes, 1977)	1.59 g glucose/g
Proportion of cross-sectional area of pine needles represented by epithelial cells and resin duct walls	0.045
Cost of construction of resin-producing cells and storage ducts in pine foliage	1.59 g glucose/g x 0.045 = 72 mg glucose/g
Proportion of monoterpenes to other constituents in pine resin	0.5
Cost of monoterpene biosynthetic machinery and storage sites in pine foliage	72 mg glucose/g x 0.5 = 36 mg glucose/g

Figure 2   Calculation of costs for biosynthesis and storage of monoterpenes.

negligible because pine foliar monoterpenes are produced in the epithelial cells lining the duct lumen and are secreted directly into the lumen. There is no evidence that this secretion requires a substantial energy input. When resin ducts are severed by herbivore attack, accumulated monoterpenes can flow long distances to the wound site (Cook and Hain, 1986; Popp *et al.*, 1991), but this movement is simply a consequence of pressure gradients and is not energetically expensive. Maintenance of stored monoterpene pools is also not costly because these pools are not subject to significant turnover or volatilization. When a pulse of $^{14}CO_2$ was applied to saplings of *Pinus contorta* and incorporation into monoterpenes monitored over 14 days, no monoterpene turnover was detected (Gershenzon *et al.*, 1993). Volatilization is also of minor significance because, under natural conditions, rates of monoterpene emission from foliage are too low to affect pool sizes (Lerdau *et al.*, 1994b). The amount of foregone photosynthesis can be estimated from the needle volume that is used for monoterpene synthesis and storage rather than photosynthetic tissue. This volume is small, and the foregone photosynthetic tissue is quite small compared to the volume of the mesophyll.

The cost formulas described above have been applied to the foliar monoterpenes of *Pinus ponderosa*, with the monoterpene concentration data coming from Lerdau *et al.* (1994b). The average concentration of monoterpenes in *Pinus ponderosa* foliage is 7.85 mg per gram fresh weight (Lerdau *et al.*, 1994b). From the calculations shown in Fig. 1, a net substrate and cofactor cost of 3.54 g glucose per gram monoterpene times 7.85 mg monoterpene per gram tissue yields a total synthesis cost of 27.8 mg glucose per gram fresh weight for the biosynthesis cost of monoterpenes. The biosynthetic machinery and storage costs are 36 mg glucose per gram fresh weight of tissue (Fig. 2). Thus, the biosynthetic costs of monoterpenes are less than half of the total cost (neglecting transport, maintenance, and foregone photosynthesis). These results indicate that the costs of making a defensive compound that requires specialized storage structures may be very low in relation to the costs of the structures themselves. This possibility, without any accompanying biochemical cost calculations, was proposed by Bjorkman *et al.* (1991) as an explanation for their results in which reduced nitrogen availability led to reductions in monoterpene concentrations. If the main costs of monoterpenes lie in their storage, then it is inappropriate to consider them as "carbon-based defenses" *sensu* Bryant *et al.* (1983) or Herms and Mattson (1992). Indeed, these results suggest that the entire distinction of carbon- versus nitrogen-based defenses may be inappropriate because it considers only substrate costs and not the full costs of defensive compounds.

These high storage costs also indicate that monoterpenes, even though they can be metabolized (Gershenzon *et al.*, 1993), are a relatively immobile defense; the resin canals cannot be broken down. Theories of chemical defense that divide defense compounds into mobile and immobile types must consider the relative mobility of the different aspects of the defenses. Just as substrate composition does not consider the most costly aspect of monoterpenes, so an analysis of mobility that only looks at monoterpene lability without considering resin canals misses the largest costs.

III. Full Costs and Constitutive versus Induced Defenses

One use of full cost calculations is to address the question of constitutive versus induced defenses. Constitutive defenses are those made by a plant prior to any signal caused by herbivore or pathogen attack. Induced defenses are synthesized in response to such an attack. Induction of defenses has been the object of much theoretical and empirical research, perhaps because it is one of the most obvious ways in which plants actively respond to their environment (Lerdau *et al.*, 1994a). There has been a good deal of theoretical discussion as to when it is "better" for a plant to use constitutive and when it is better to use induced defenses. Karban and Myers (1989) recognize that the frequency and predictability of herbivory goes a long toward explaining why certain plants appeared to rely on induced and others on constitutive defenses. As with many issues in defense allocation theory, modeling efforts have been divorced from empirical research because the models require knowing costs, and full costs are difficult to measure.

Monoterpenes are an ideal set of compounds with which to examine the circumstances favoring constitutive and induced defenses because (1) full cost calculations can be done and (2) these compounds exist in plants as both constitutive and inducible defenses (Gershenzon and Croteau, 1993). The high cost of storage structures provides an interesting twist because, even in species that induce the synthesis of new monoterpenes, the production of storage structures is constitutive. Constitutive monoterpenes accumulate exclusively in storage structures, whereas uninduced monoterpenes accumulate either in existing storage structures or within wounded tissue itself, in large cavities outside differentiated storage structures (Cheniclet *et al.*, 1988; Lewinsohn *et al.*, 1992). Thus, one can use the allocation to storage structures (resin canals) as a surrogate for the relative importance of constitutive defenses. Matson and Hain (1987) suggest that southeastern pine species, native to regions without severe winters, face more constant insect pressure than pines native to western mountains. They go on to show that southeastern pine species are more likely to have extensively developed foliar resin canals than are pine species native to the intermountain West (see Table 2 of Matson and Hain, 1987). These results, though indirect, suggest that, within the genus *Pinus* in North America, the relative importance of constitutive versus induced defense depends on the regularity of herbivore attack.

Other factors may explain the differences in relative inducibility between the warm- and cold-winter populations of *Pinus*. However, cortical monoterpenes in conifers are linked very closely to chemical defense rather than to other physiological functions such as reduction of water loss (as has been found for monoterpenes stored on the surfaces of leaves; Mooney *et al.*, 1980). These results should be taken as highly suggestive rather than as definitive proof of the role of herbivores in affecting terpene allocation patterns and relative inducibility.

IV. Summary

The broad range of theories concerning plant allocation to defense are all explicitly evolutionary in their underlying assumptions. Those theories based on life history strategy that use seed production as their currency are sometimes considered as evolutionary theories, with those theories that use material quantities such as biomass for currency considered physiological. However, this distinction is artificial; all the theories use optimality principles in calculating the costs and benefits of resource allocation strategies. The theories differ in the questions they can be used to address, the taxa to which they can be applied, and the time and taxonomic scales over which they operate.

Monoterpenes are a class of defensive compounds whose biochemistry is particularly well understood. This understanding has allowed calculation of the full cost of allocation. Contrary to calculations based on incomplete cost estimates, the substrate and cofactor costs are only approximately 40% of the real cost of monoterpene production. The storage structures make up over 50% of the full costs. This importance of storage structures suggests that distinctions among defensive compounds based on substrate composition (e.g., carbon- versus nitrogen-based defenses) are inappropriate when studying the costs of allocation to defense.

Monoterpenes are also useful for studying induced and constitutive defenses. Within the genus *Pinus* in North America, the relative importance of monoterpenes as induced and constitutive defenses varies with the frequency of herbivore attacks. This variation can be indexed by measuring the relative abundance of storage structures (resin canals), a constitutive allocation even in those species that induce monoterpene synthesis in response to attack. This indexing is valid because of the large cost of the resin canals relative to the costs of the monoterpenes themselves. A comparative study of foliar resin canals in North American *Pinus* species suggests that regularity of insect attack correlates with the relative importance of allocation to constitutive defenses. These results indicate the utility of full cost analyses in addressing large-scale ecological questions.

References

Banthorpe, D., and Charlwood, B. (1980). The terpenoids. *In* "Encyclopedia of Plant Physiology" (E. Bell, and B. Charlwood, eds.), Vol. 12, pp. 185–219. Springer-Verlag, Berlin.

Bernard-Dagan, C., Carde, J., and Gleizes, M. (1979). Etude des composés terpénique au cours de la craoissance des aiguilles du Pin maritime: Comparaison de données biochimiques et ultrastructurales. *Can. J. Bot.* **57,** 255–263.

Bjorkman, C., Larsson, S., and Gref, R. (1991). Effects of nitrogen fertilization on needle chemistry and sawfly performance. *Oecologia* **86,** 202–209.

Bryant, J., Chapin III, F., and Klein, D. (1983). Carbon/nutrient balance of boreal plants in relation to vertebrate herbivory. *Oikos* **40,** 357–368.

Cheniclet, C., Bernard-Dagan, C., and Pauly, G. (1988). Terpene biosynthesis under pathological conditions. *In* "Mechanisms of Woody Plant Defense against Insects" (W. Mattson, J. Levieux, and C. Bernard-Dagan, eds.), pp. 117–130. Springer-Verlag, New York.

Chung, H., and Barnes, R. (1977). Photosynthate allocation in *Pinus taeda*. I. Substrate requirements for synthesis of shoot biomass. *Can. J. For. Res.* **7,** 106–111.

Cohen, D. (1966). Optimizing reproduction in a randomly varying environment. *J. Theor. Biol.* **12,** 10–129.

Coley, P., Bryant, J., and Chapin III, F. (1985). Resource availability and plant anti-herbivore defense. *Science* **230,** 895–899.

Cook, S., and Hain, F. (1986). Defensive mechanisms of loblolly and shortleaf pine against attack by southern pine beetle, *Dendroctonus frontalis* Zimmerman, and its fungal associate, *Caretocystis minor* (Hedgecock) Hunt. *J. Chem. Ecol.* **12,** 1397–1406.

Croteau, R., and Johnson, M. (1985). Biosynthesis of terpenoid wood extractives. *In* "Biosynthesis and Biodegradation of Wood Components" (T. Higuchi, ed.), pp. 379–439, Academic Press, New York.

Fahn, A. (1979) "Secretory Tissues in Plants." Academic Press, London.

Fahn, A., and Benayoun, J. (1976). Ultrastructure of resin ducts in *Pinus halepensis. Ann. Bot.* **40,** 857–863.

Feeny, P. (1976). Plant apparency and chemical defense. *In* "Biochemical Interactions between Plants and Insects" (J. Wallace, and R. Mansell, eds.), pp. 1–40. Plenum, New York.

Gershenzon, J. (1994a). Metabolic costs of terpenoid accumulation in higher plants. *J. Chem. Ecol.* **20,** 1281–1328.

Gershenzon, J. (1994b). The cost of plant chemical defense against herbivory: A biochemical perspective. *In* "Insect–Plant Interactions" (E. Bernays, ed.), Vol. 5, pp. 105–173. CRC Press, Boca Raton, Florida.

Gershenzon, J., and Croteau, R. (1993). Terpenoid biosynthesis: The basic pathway and formation of monoterpenes, sesquiterpenes, and diterpenes. *In* "Lipid Metabolism in Plants" (T. Moore, ed.), pp. 333–388. CRC Press, Boca Raton, Florida.

Gershenzon, J., Murtagh, G., and Croteau, R. (1993). Absence of rapid terpene turnover in several diverse species of terpene-accumulating plants. *Oecologia* **96,** 583–592.

Harborne, J., and Tomas-Barberan, T., eds. (1991). "Ecological Chemistry and Biochemistry of Plant Terpenoids," Annual Proceedings of the Phytochemical Society of Europe, Vol. 31. Oxford Univ. Press (Clarendon), Oxford.

Herms, D., and Mattson, W. (1992). The dilemna of plants: To grow or defend. *Q. Rev. Biol.* **67,** 283–335.

Karban, R., and Myers, J. (1989). Induced plant responses to herbivory. *Annu. Rev. Ecol. Syst.* **20,** 331–348.

Koide, R., and Elliot, G. (1989). Cost, benefit, and efficiency of the vesicular–abuscular mycorrhizal symbiosis. *Funct. Ecol.* **3,** 252–255.

Krebs, J., and McCleery, R. (1984). Optimization in behavioural ecology. *In* "Behavioural Ecology" (J. Krebs, and N. Davies, eds.), pp. 91–121. Blackwell, London.

Langenheim, J. (1994). Higher plant terpenoids: A phytocentric overview of their ecological roles. *J. Chem. Ecol.* **20,** 1223–1280.

Lerdau, M. (1992). Future discounts and resource allocation in plants. *Funct. Ecol.* **6,** 371–375.

Lerdau, M. (1993). Formal equivalence among resource allocation models: What is the appropriate currency? *Funct. Ecol.* **7,** 507–508.

Lerdau, M., Litvak, M., and Monson, R. (1994a). Monoterpenes and theories of plant chemical defense. *Trends Ecol. Evol.* **9,** 58–61.

Lerdau, M., Dilts, S., Lamb, B., Westberg, H., and Allwine, E. (1994b). Monoterpene emission from ponderosa pine. *J. Geophys. Res.* **99**(D), 16609–16615.

Lewinsohn, E., Gijzen, M., and Croteau, R. (1992). Regulation of monoterpene biosynthesis in conifer defense. *In* "Regulation of isopentenoid metabolism." *ACS Symp. Ser.* **497,** 8–17.

Lorio, P. (1986). Growth–differentiation balance: A basis for understanding southern pine beetle–tree interactions. *For. Ecol. Manage.* **14,** 259–273.

McKey, D. (1979). The distribution of secondary compounds within plants. *In* "Herbivores: Their Interactions with Secondary Plant Metabolites" (G. Rosenthal, and D. Janzen, eds.), pp. 55–133. Academic Press, Orlando, Florida.

Matson, P., and Hain, F. (1987). Host conifer defense strategies: A hypothesis. *In* "Proceedings of the First IUFRO Conference." Banff, Canada.

Merino, J., Field, C., and Mooney, H. (1984). Construction and maintenance costs of Mediterranean-climate evergreen and deciduous leaves. II. Biochemical pathway analysis. *Oecol. Plant.* **5**, 211–229.

Mooney, H. (1972). The carbon balance of plants. *Annu. Rev. Ecol. Syst.* **3**, 315–346.

Mooney, H., Ehrlich, P., Lincoln, D., and Williams, K. (1980). Environmental controls on the seasonality of a drought-deciduous shrub, *Diplacus aurantiacus*, and its predator, the checkerspot butterfly, *Euphydryas chalcedona. Oecologia* **145**, 143–146.

Penning de Vries, F., Brunsting, A., and Van Laar, H. (1974). Products, requirements, and efficiency of biosynthesis: A quantitative approach. *J. Theor. Biol.* **45**, 339–377.

Perrin, N. (1993). On future discounts and economic analogy in life-history studies. *Funct. Ecol.* **7**, 506–507.

Popp, M., and Johnson, J., and Massey, T. (1991). Stimulation of resin flow in slash and loblolly pine by bark beetle vectored fungi. *Can. J. For. Res.* **21**, 1124–1126.

Rhoades, D. (1979). Evolution of plant chemical defense against herbivores. *In* "Herbivores: Their Interactions with Secondary Plant Metabolites" (G. Rosenthal, and D. Janzen, eds.), pp. 3–54. Academic Press, Orlando.

Rhoades, D., and Cates, R. (1976). Toward a general theory of plant antiherbivore chemistry. *Rec. Adv. Phytochem.* **10**, 168–213.

Rothschild, M., and Stiglitz, J. (1970). Increasing risk I: A definition. *J. Econ. Theory* **2**, 224–243.

Simms, E., and Rausher, M. (1987). Costs and benefits of plant defense to herbivory. *Am. Nat.* **130**, 570–581.

Stearns, S. (1989). Trade-offs in life-history evolution. *Funct. Ecol.* **3**, 259–268.

Taylor, H., Gourley, R., Lawrence, C., and Kaplan, R. (1974). Natural selection and life history attributes: An analytical approach. *Theoretical Population Biology* **5**, 104–122.

Werker, E., and Fahn, A. (1969). Resin ducts of *Pinus halepensis* Mill. Their structure, development, and pattern of arrangement. *Bot. J. Linn. Soc.* **62**, 379–411.

Zangerl, A., and Bazzaz, F. (1992). Theory and pattern in plant defense allocation. *In* "Plant Resistance to Herbivores and Pathogens" (R. Fritz and E. Simms, eds.). Univ. of Chicago Press, Chicago.

12

Toward Models of Resource Allocation by Plants

John Grace

I. Introduction

There is an urgent need for simple models that enable the researcher to simulate plant growth in monocultures or mixtures, in relation to natural and man-made variations in resource availablity that occur from time to time. Such models are currently needed for predicting the spread of species in relation to such global changes as elevated CO_2, nitrogen deposition, and the incidence of drought; or for allocating nitrogen to the canopy to simulate fluxes of carbon dioxide and water (Lloyd *et al.*, 1995; Lloyd and Farquhar, 1996; Williams *et al.*, 1996). Growth is the summation of many processes, some of which are understood better than others. Of these processes, photosynthesis and respiration are easily measured and relatively well understood by biochemists and physiologists; well-established mathematical representations of these processes for leaves and ecosystems are available (e.g., von Caemmerer and Farquhar, 1981; Lloyd *et al.*, 1995). In contrast, allocation of carbon or nitrogen to shoot versus root, although just as important as photosynthesis as a determinant of growth, is still poorly understood. Progress is hampered most of all by the lack of convenient methodology to measure directly the allocation of carbon and nitrogen, quantitatively and on a continuous basis. Despite the enthusiasm for tracer techniques that occurred 20 and 30 years ago, interpretational difficulties seemed to have overwhelmed such studies, and most recent workers in the allocation field resort to destructive sampling, despite its poor temporal resolution.

Plants achieve their structure and perpetuate themselves by concentrating elemental resources from very dilute surroundings. This requires an energy supply to do work, and in vascular plants it requires a set of suitably proportioned peripheral organs (leaves and roots) to acquire resources and a transport system to deliver the materials to the appropriate destination. At the coarsest level of enquiry, we may ask how the plant "decides" on its allocation between leaf, stem, and root. One response, found in even the earliest work on the subject (e.g., Brenchley, 1916; Blackman, 1919; Brouwer, 1962), is that the plant is a "wise investor" of resources. According to this view, the plant most capable of survival in the struggle for existence was the one that invests resources in such a way that it protects itself from adversity and defends itself against its enemies, and grows the fastest. It is therefore able successfully to complete its life cycle in the face of competition with neighbors. In fertile environments, where nutrients and water may not be limiting, the most successful plant is expected to be the one making a large investment in leaves, thus capturing energy and producing much glucose to "spend" on the synthesis of biochemical constituents. In nutrient-poor or dry soils, however, the attempt to acquire energy by leaf growth needs to be tempered by the need to develop proportionately more surface area below the ground to acquire resources from within the soil, so root growth is required. As Brenchley (1916) wrote, "the plant makes every endeavor to supply itself with adequate nutrients and . . . when the food supply is low, it strives to make as much root as possible."

In both resource-rich and resource-poor environments it is necessary for a plant to compete with neighbors for the space within which its organs (leaves and roots) are actively absorbing carbon dioxide, nutrients, and water. In practice, plants develop a root:shoot ratio that is partly inherited and partly adjusted over time scales of days to years according to the relative concentrations in above- and belowground resources. Plants are believed to sense the external environment and respond to fluctuations in resource availablility by applying physiological controls that alter development (Aphalo and Ballaré, 1995). An example of such a sensing mechanism is the phytochrome system that provides information on the presence of competing plants. It does this by detecting the shade-light in the red and far red regions of the spectrum (Ballaré et al., 1995). The response of the plant varies with species, but ruderals typically respond to the shade-light cast by other plants by allocating resources to height growth, thus overtopping competitors. However, few such mechanisms of environmental sensing by plants are so well characterized and understood. The representation of these imperfectly understood external sensing mechanisms in models of plant growth is unsatisfactory, as it leads to models with an underlying structure that may be disputed, as well as uncertain parameters, many of which are determined with difficulty or simply guessed.

II. Functional Equilibrium Models

Problems associated with the uncertainties in models that are realistic, and therefore highly complex, are often circumvented by models of allocation which are parsimonious, and in which no external sensing mechanism is required, as in the so-called functional equilibrium models (Wilson, 1988). An example is the diffusion resistance model of Thornley (1972a,b). In Thornley's model, the uptake rate of nitrogen and carbon dioxide increases with external concentration up to a point where the uptake mechanism becomes saturated. The N and C taken up enter biochemical pools in the roots, stem, and shoots. Then, root, stem, and leaf meristems are allowed to grow by consuming the N and C present in the pools, according to the carbon:nitrogen ratio that they require to satisfy their "standard" chemical composition. Replenishment from the soluble pools occurs by diffusion through transport resistance which connects the root and shoot pools; thus, nitrogen diffuses upward from the root pools (where it is relatively concentrated) and carbon diffuses downward to the root pools (where it is relatively dilute). This model works well in the sense of adjusting the root:shoot ratio in relation to shade or nutrient shortage, for example. However, root:shoot ratios in nature are found to depend not only on irradiance and nutrition, but also on the supply of water. Plants grown in dry soil acclimate in several ways, one of which is the development of proportionately more root, enabling them to tolerate drought (e.g., Khalil and Grace, 1992). The mechanism of this response is not only a matter of allocation, but involves stomatal regulation. Sensing of water shortage is thought to include movement of abscisic acid in the transpiration stream, as a chemical messenger, and a role for abscisic acid in causing some degree of stomatal closure is proposed as well (Davies and Zhang, 1991; Khalil and Grace, 1993). Models of this mechanism are provisional, because knowledge is incomplete; yet it is vital to incorporate a representation of the response to soil water shortage into allocation models. Dewar (1993) has extended Thornley's model to include an allocation response to water, whereby an increased allocation to roots occurs when water is in short supply.

Although Thornley's model is not intended to represent physiological or biochemical mechanisms, its basic assumptions are consistent with the classic observations of phloem transport: movement of sugars occurs from soluble pools of high concentration in the mature leaves to pools of low concentration in actively growing regions such as the young shooots and the roots. The real process is too fast, and too responsive to temperature, to be ascribed to physical diffusion. One of the earliest theories of phloem transport, the Münch hypothesis, proposed that the sugars were swept along by an osmotically driven mass flow in the phloem cells; however, this is now thought most unlikely because the physical resistance to flow would be too great. In fact, there is no real consensus on the mechanism of phloem transport, despite over a century of active research and a bewildering array of possible mechanisms (Zimmermann and Milburn, 1975; Moorby, 1981). Dewar (1993) draws attention to the need for models to reflect, as far as possible, physiological understanding of the processes, and points out the consistency of his own model with the Münch hypothesis. Until mechanisms are more fully understood, the only test of the model can be the extent to which it mimics the general behavior of real plants, which is to allocate to root or shoot according to whether it is the aboveground or belowground resources that are in short supply. Often the simplest models have great intellectual appeal, because of their mathematical elegance, and they often pass this first test of realistic behavior. We should, however, keep in mind the caveat that models based on an incorrect view of the underlying mechanisms will often generate results that look correct, especially when they have been constructed *a posteriori*, but they may fail in predicting the behavior of the system in new and unchartered situations.

III. Can the Most Simple Models Be Useful?

In this chapter we present a "toy" model that contains one of the simplest possible sets of relationships to permit the modeling of plant growth in relation to the supply of carbon and nutrients. It supposes only that the plant senses its own elemental composition, and responds by attempting to achieve internal chemical homeostasis.

The elemental composition of leaves varies somewhat, but it is not unreasonable to assume that for a given species there is an optimal composition, which we think of in this model as the target composition. In the "toy" version of the model, it will not be necessary to represent all the essential elements. Let us, for demonstration purposes, assume that the plant normally has n carbon atoms to each nitrogen atom, and that allocation is driven by the need to maintain a constant n. Thus, when there is insufficient nitrogen, allocation is directed to roots, and when there is insufficient carbon, allocation is directed to the production of leaves (Fig. 1). The computation rules are thus

$$\text{IF } C/N > n \text{ THEN (grow root)}$$

$$\text{ELSE (grow shoot)}$$

and root and shoot growth is achieved by adding the incoming nitrogen and carbon to the existing root or shoot.

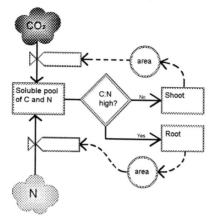

Figure 1 Simple flow diagram for modeling the distribution of C and N in relation to the C:N ratio. Solid lines denote flows of material, and broken lines mean influences.

Following Thornley (1972a,b), the rate of uptake of C and N is determined using a rectangular hyperbola to describe the dependency of the flux of carbon and nitrogen on their respective concentrations:

$$F_j = \frac{F_{\text{max}j}X_j}{a_j + X_j} \times A_j,$$

Where $F_{\text{max}j}$ is the maximum flux of carbon per leaf area ($j = 1$) or nitrogen per root area ($j = 2$), a_j is that concentration of C in air (as CO_2) or N in soil (as nitrate or ammonium) which gives half the maximum flux, and A_j is the area of root or shoot obtained by multiplying the current biomass of the root or shoot by a constant b_j. Thus F_1 corresponds to what is normally called the net assimilation rate of leaves, and F_2 is the equivalent quantity applied to roots. In the following demonstration of the model, we have taken arbitrary values to illustrate the principle. For carbon, $F_{\text{max}1} = 0.1$, $a_1 = 2.0$, $b_1 = 5.0$; and for nitrogen, $F_{\text{max}2} = 0.1$, $a_2 = 2.0$, $b_2 = 10.0$. It is further assumed in this demonstration that external concentrations of resources remain constant, as they would in well-ventilated plants growing in continuously flowing hydroponic solutions.

In practice, such a model oscillates between root and shoot growth. When roots grow, the increasing area takes up more N until that element is in surplus. Then root growth stops and all C and N is directed to the shoots until C/N exceeds its critical value.

After 100 iterations, representing days, the root:shoot ratio stabilized at 1.86, whereas the biomass increases exponentially (Fig. 2). If we double the CO_2 concentration while keeping the nitrogen constant, we achieve a

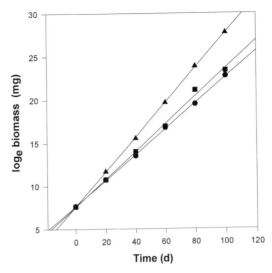

Figure 2 Exponential growth, as estimated from the model, under present-day conditions (●), twice normal CO_2 (■), and twice normal N concentrations around the roots (▲). The root:shoot ratio in these runs stabilized at 1.86, 1.95, and 1.01, respectively.

modest increase in biomass and an upward shift in root:shoot ratio from 1.3 to 1.95 as the plant diverts more carbon and nitrogen to roots, enabling more N to be taken up. The increase is only modest, because the CO_2 concentration is near-saturating in this example. However, when we double the N concentration at normal CO_2, the rate of increase in biomass is greater, as the N concentrations used in this example are nowhere near saturating. Moreover, the root:shoot ratio shifts downward from 1.3 to 1.0 as the plant must now take up more C to match its N and thus maintain the C:N ratio.

We now rerun the model, to extend the range of N concentration from 0.001 to 0.64 kg m^{-3}. After 100 days the biomass has responded strongly to N concentration until it reaches about 0.04 (log 0.04 = −3.21) and is thereafter less responsive (Fig. 3). Over the entire range, the root:shoot ratio declines remarkably from over 3.0 to less than 0.25 as the concentration of N increases from 0.005 to 0.64. The pattern is modified by simulating a shade treatment, achieved by reducing the carbon flux to 0.025 of its calculated value at each iteration. Under shade, we see that the model allocates to leaf, but the overall effect of increasing N is still to cause a decline in the root:shoot ratio (Fig. 3).

It is clear that the response of plant growth and root:shoot ratios to changing environmental variables, as shown in this model, is consistent with the typical responses shown by most plants (Hunt, 1978; MacDonald

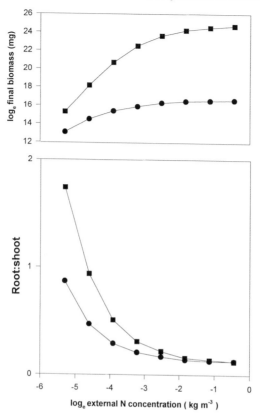

Figure 3 Influence of varying the nitrogen supply, expressed as concentration around the roots, on the biomass at 100 days and on the root:shoot ratio. The runs were performed at the full rate of photosynthesis (■) and were repeated with photosynthesis set at one-quarter of the normal rate (●), to simulate the effect of shade.

et al., 1986; Wilson, 1988; Wikström and Ericsson, 1996). The model shows a very similar response pattern to the Thornley (1972a,b) model but is simpler in construction, and simple enough to be useful as a basis for models of geographical spread, for example, when expressed as a cellular automa (Colasanti and Grime, 1993), or for a succession model, after the addition of "vital attributes" or characters that describe the life history and habit of the species.

Objections may be raised to the use of the C:N ratio as a control variable, both here and in the Thorney model. For one thing, no mechanisms whereby plants respond to gross elemental ratios is known, and it is difficult to conceive of one. Second, C:N is not a constant for any species, not even within any one organ. There are rather few good data sets of the whole-plant and whole-organ C:N ratio of species grown under a wide range of conditions, but there are several where the C:N ratio of leaves has been reported. Wong *et al.* (1992) grew four *Eucalyptus* species at two nitrogen concentrations and two CO_2 concentrations, and a similar data set is available for *Pinus* in Griffin *et al.* (1996). The published data suggest that leaf C:N does vary considerably on a percent leaf mass basis. As expected, elevated CO_2 results in a "dilution" of the N, but the quantity C:N is much more conservative when reexpressed on a whole-plant basis, particularly if storage carbohydrates are excluded from the calculation (S. C. Wong and J. Lloyd, personal communication). The reason for the conservative nature of ratios such as C:N or C:P is that cytoplasm is rather constant in elemental composition, and the main variation is introduced by the relatively carbon-rich structural components cellulose, lignin, and, in some species, waxes and oils. The extent of cellulose and lignin does not change very much in herbaceous plants. In marine organisms, many researchers have considered that the elemental composition of biomass is sufficiently fixed to be representable as $C_{106}N_{16}P$, known as the Redfield formula (Redfield, 1958; Falkowski and Woodhead, 1992). In vascular plants the C:N ratio is likely to vary greatly between wood and leaf, and hence a terrestrial Redpath formula is unlikely to be useful.

IV. Rules for Allocation to Wood

This leads us to consider the case of trees, where mechanical properties are very important. On a macroscale a tree must resist the forces of gravity and wind. This requires an investment of carbon in structural tissue (massive roots, stems, and branches). It has been observed many times that allocation patterns to wood are adaptive, such that wood is formed in the places where it is most needed for safety. This phenomenon is best seen in the comparison of staked and unstaked trees (Larson, 1965), in the response of trees to mechanical stimulation (Valinger, 1993), and in the keenly observed sketches of mechanically challenged trees by Mattheck (1991). As for the sensing mechanism, it is clear that plants can detect mechanical stimuli, and the term thigmomorphogenesis was coined by Jaffe (1973) to summarize the process whereby development of form responds to mechanical stimulation. However, the phenomenon is not understood, and it has hitherto been ignored in allocation models of trees. One possible method to calculate the allocation required for mechanical support would be to simply calculate the forces that would act on the tree in an exceptional gale using the methods outlined in Woods (1995), and then ask, Is there enough investment in structural tissue to confer an appropriate margin of safety? If the answer is "no", carbon would be diverted to the required locations of stem and branches. Unfortunately, this calculation is not easily done, as much depends on the dynamic nature of the action of wind on the tree, the damping of oscillations by neighboring trees, and the condition of the soil when the gale strikes. In fact, trees are more often uprooted rather than snapped off, and allocation to structural roots is probably of paramount importance.

Stems and branches have a dual role. As well as being important for structural support of the crown, they are also important as transport pathways for water. On a microscale, the individual tracheids and vessel elements must be very stiff to avoid collapse under the great tensions in the water when the tree is actively transpiring. On a macroscale, the importance of proper hydraulic design was evident to Leonardo da Vinci. He observed that "all the branches of a tree at every stage of its height, when put together, are equal in thickness to the trunk below them" (Richter, 1970). A strict proportionality of leaf area and sapwood area is generally observed in trees and shrubs (Shinozaki *et al.*, 1964; Waring *et al.*, 1982; Valentine, 1985). The overall theory relating the evaporative demand for water to the hydraulic pathway in the plant is the pipe model paradigm. In this theory, a unit of transpiring area (leaf) is necessarily equipped with a unit of water-transporting pipework (xylem). If the xylem quantity falls below the appropriate amount, the hydraulic resistance is too great and the leaf consequently suffers from water shortage; and if the xylem is excessive, this represents a wasteful allocation of carbon to nonproductive tissue. The theory has great intellectual appeal, and may be a useful basis for the allocation algorithm in models of tree growth. However, as is often the case, close examination shows that it is an oversimplification. When the ratio of leaf area to sapwood is examined over a wide geographical range, it is found not to be constant. Mencuccini and Grace (1995) found that Scots pine grown in a dry part of England had proportionately less leaf area than trees from the same seed source grown in the wetter climate of Scotland, and that water potentials measured in the canopy were no more negative in the dryer region. It seems that allocation is adjusted over the life of the tree, so that the proportions of leaf to wood are commensurate with the evapotranspirational demand imposed by the environment. Moreover, the leaf area:sapwood ratio changes systematically with the age and size of the tree, so that a tall tree has proportionately less leaf area, presumably in response to the additional hydraulic resistance imposed by the increased path length, and by the gravitational contribution to the water potential. Indeed, it has been suggested that the height growth of trees is ultimately limited by the difficulty in raising water to such great height.

If there is insufficient water-conducting tissue, then the tensions in the water columns may become so great during active transpiration that cavitation of water occurs, and a proportion of the conducting elements may become dysfunctional. According to the runway cavitation hypothesis (Tyree and Sperry, 1989), the consequent loss of conductivity could cause further cavitation, leading to catastrophic rates and a cessation of water supply altogether. However, it seems likely that the highest hydraulic resistances are in the twigs, so that they would be the first part of the tree to suffer from cavitation; thus, the foliage area would be reduced, and the leaf area to sapwood area would be adjusted. These considerations have led to the construction of a model, HYDRALL, which allocates carbon on the basis of the water potential in the canopy, in such a way that cavitation is avoided (Magnani and Grace, 1997). The model uses a soil–vegetation– atmosphere transfer scheme (SVATS) to calculate the leaf water potentials in relation to the fluctuating atmospheric conditions. When the diurnal cycle of water potential falls to the critical point at which the species begins to cavitate (Tyree and Sperry, 1989), carbon allocation is diverted away from the canopy to the sapwood or fine roots, whichever has the lowest marginal cost of production. Using only this principle, the model behaves as the plant behaves, adjusting the leaf area to sapwood ratio in accordance with the evaporative conditions and with the age of the tree, and the growth rate of the tree declines with age just as it does in life.

At this stage, it seems that we have a good knowledge of which variables influence the patterns of allocation in both woody and herbaceous plants, and we have a wealth of experimental data from which we can assess the magnitude of such effects. We also have a set of concepts which enable us to build models of plant growth that are useful in testing our understanding, and even useful in a practical way to predict plant growth in a changing environment (Thornley and Cannell, 1992).

V. Remaining Difficulties

Rather little progress has been made toward understanding *priorities.* Given that a tree, in order to survive, needs to develop a structure that can acquire scarce resources and withstand mechanical stress, what will it do when faced with a conflicting need? For example, in a fertile site a tree should develop leaf area rather than root system, but if the site happens to be windy this line of development would lead to premature uprooting in gale force winds. Moreover, there are complex interactions that models ought to take into account. For example, an elevated concentration of CO_2 may first lead to an increase in the allocation to root exudates, which may lead to a stimulation in the soil organisms, which in turn increases mineralization (Zak *et al.,* 1993). The interaction between water and nutrients is still not well understood. For instance, the hydraulic conductivity of the root system is dependent on the nutrient supply (Radin and Mathews,

1989; Carvajal *et al.*, 1996), so that nutrient-starved plants may often be water stressed as well.

We have already pointed out that the use of models based on imperfectly understood mechanisms, or incompletely represented mechanisms, is likely to mislead. All the models discussed so far are based on the notion of the "wise allocator of resources," that is, they use optimal allocation theory. The underlying assumption here is that the evolutionary process has weeded out all patterns that are not optimized. However, as evolution is a continuing process it may be dangerous to make this assumption, for we may be grossly overestimating the capacity of plants to acclimate. In today's changing chemical and physical climate, we can expect to face new evolutionary challenges as new combinations of chemical and physical variables appear. One example relates to the allocation of mass in relation to magnesium supply (Wilkström and Ericsson, 1995). For the case of this nutrient, a decline in applied concentration causes an allocation to shoot instead of root, hence increasing the impact of any Mg shortage. In cases where the supply of N relative to that of Mg has increased, because of anthropogenic influences, this response may have caused Mg deficiencies and tree decline.

References

Aphalo, P. J., and Ballaré, C. L. (1995). On the importance of information-acquiring systems in plant–plant interactions. *Funct. Ecol.* **9,** 5–14.

Ballaré, C. L., Scopel, A. L., Roush, M. L., and Radosevich, S. R. (1995). How plants find light in patchy environments: A comparison between wild-type and phytochrome-B deficient mutant plants of cucumber. *Funct. Ecol.* **9,** 859–868.

Blackman, V. H. (1919). The compound interest law and plant growth. *Ann. Bot.* **33,** 353–360.

Brenchley, W. E. (1916). The effect of concentration of the nutrient solution on the growth of barley and wheat in water culture. *Ann. Bot.* **30,** 77–90.

Brouwer, R. (1962). Distribution of dry matter in the plant. *Neth. J. Agric. Sci.* **10,** 399–408.

Carvajal, M., Cooke, D. T., and Clarkson, D. T. (1996). Responses of wheat plants to nutrient deprivation may involve the regulation of water-channel function. *Planta* **199,** 372–381.

Colasanti, R. L., and Grime, J. P. (1993). Resource dynamics and vegetation processes: A deterministic model using 2-dimensional cellular automata. *Funct. Ecol.* **7,** 169–176.

Davies, W. J., and Zhang, J. (1991). Root signals and the regulation of growth and development of plants in drying soils. *Annu. Rev. Plant Physiol. Plant Mol. Biol.* **42,** 55–76.

Dewar, R. C. (1993). A root–shoot partitioning model based on carbon, nitrogen and water interaction and the Münch phloem flow. *Funct. Ecol.* **7,** 356–368.

Falkowski, P. G., and Woodhead, A. D. (1992). "Primary Production and Biogeochemical Cycles in the Sea." Plenum, New York.

Griffin, K. L., Winner, W. E., and Strain, B. R. (1996). Construction costs of loblolly and ponderosa pine leaves grown with varying carbon and nitrogen availability. *Plant Cell Environ.* **19,** 729–739.

Heilmeier, H., Erhard, M., and Schulze, E.-D. (1997). Biomass allocation and water use under arid conditions. This volume, Chapter 4.

Hunt, R. (1978). "Plant Growth Analysis." Arnold, London.

Jaffe, M. J. (1973). Thigmomorphogenesis: The response of plant growth and development to mechanical stimulation. *Planta* **114**, 143–157.

Khalil, A. A. M., and Grace, J. (1992). Acclimation to drought in *Acer pseudoplatanus* seedlings. *J. Exp. Bot.* **257**, 1591–1602.

Khalil, A. A. M., and Grace, J. (1993). Does xylem ABA control the stomatal behaviour of water-stressed sycamore (*Acer pseudoplatanus* L.) seedlings? *J. Exp. Bot.* **44**, 1127–1134.

Larson, P. R. (1965). Stem form of young *Larix* as influenced by wind and pruning. *For. Sci.* **11**, 412–424.

Lloyd, J., and Farquhar, G. D. (1996). The CO_2 dependency of photosynthesis, plant growth response to elevated CO_2 concentrations and their interaction with soil nutrient status. 1. General principles and forest ecosystems. *Funct. Ecol.* **10**, 4–32.

Lloyd, J., Grace, J., Miranda, A. C., Meir, P., Wong, S. C., Miranda, H., Wright, I., Gash, J. H. C., and McIntyre, J. (1995). A simple calibrated model of amazon rainforest productivity based on leaf biochemical properties. *Plant Cell Environ.* **18**, 1129–1145.

MacDonald, A. J. S., Lohammer, T., and Ericsson, A. (1986). Growth response to a step-wise decrease in nutrient availability in small birch (*Betula pendula* Roth.). *Plant Cell Environ.* **9**, 427–432.

Magnani, F., and Grace, J. (1996). HydrAll—Simulating optimal carbon allocation and tree growth under hydraulic constraints in Scots pine (*Pinus sylvestris* L.). Submitted for publication.

Mattheck, C. (1991). "Trees: The Mechanical Design." Springer-Verlag, Berlin.

Mencuccini, M., and Grace, J. (1995). Climate influences the leaf area–sapwood relationship in Scots pine (*Pinus sylvestris* L.). *Tree Physiol.* **15**, 1–10.

Moorby, J. (1981). "Transport Systems in Plants." Longmans, London.

Radin, J. W., and Mathews, M. A. (1989). Water transport properties of cortical cells in roots of nitrogen- and phosphorus-deficient seeedlings. *Plant Physiol.* **89**, 264–268.

Redfield, A. C. (1958). The biological control of chemical factors in the environment. *Am. Sci.* **46**, 205–221.

Richter, J. P. (1970). "The Notebooks of Leonardo da Vinci (1452–1592), Compiled and Edited from the Original Manuscripts." Dover, New York.

Shinozaki, K., Yoda, K., Hozumi, K., and Kira, T. (1964). A quantitative analysis of plant form—The pipe model theory. 1. Basic analyses. *Jpn. J. Ecol.* **14**, 97–105.

Thornley, J. H. M. (1972a). A model to describe the partitioning of photosynthate during vegetative growth. *Ann. Bot.* **36**, 419–430.

Thornley, J. H. M. (1972b). A balanced quantitative model for root : shoot ratios in vegetative plants. *Ann. Bot.* **36**, 431–441.

Thornley, J. H. M., and Cannell, M. G. R. (1992). Nitrogen relations in a forest plantation—Soil organic matter and an ecosystem model. *Ann. Bot.* **70**, 137–151.

Troughton, A. (1974). The growth and function of the root in relation to the shoot. *In* "Structure and Function of Primary Root Tissues" (J. Kolek, ed.), pp. 153–164. Slovak Academy of Sciences, Bratislava.

Tyree, M. T., and Sperry, J. S. (1989). Vulnerability of xylem to cavitation and embolism. *Annu. Rev. Plant Physiol. Mol. Biol.* **40**, 19–38.

Valentine, H. (1985). Tree-growth models: Derivations employing the pipe model theory. *J. Theo. Biol.* **117**, 579–585.

Valinger, E. (1993). Effects of wind sway on stem form and crown development of Scots pine (*Pinus sylvestris* L.). *Aust. For.* **55**, 15–21.

Von Caemmerer, S., and Farquhar, G. D. (1981). Some relationships between the biochemistry of photosynthesis and the gas exchange of leaves. *Planta* **153**, 376–387.

Waring R. H., Schroeder, P. E., and Oren, R. (1982). Application of the pipe model theory to predict canopy leaf area. *Can. J. For. Res.* **12**, 556–560.

Wikström, F., and Ericsson, T. (1995). Allocation of mass in trees subject to nitrogen and magnesium limitation. *Tree Physiol.* **15**, 339–344.

Williams, M., Rastetter, E. B., Fernandes, D. N., Goulden, M. L., Wofsy, S. C., Shaver, G. R., Mellilo, J. M., Munger, J. W., Fan, S.-M., and Nadelhoffer, K. J. (1996). Modelling the soil–plant–atmosphere continuum in a *Quercus–Acer* stand at Harvard Forest: The regulation of stomatal conductance by light, nitrogen and soil/plant hydraulic properties. *Plant Cell Environ.* **19**, 911–927.

Wilson, J. B. (1988). A review of the evidence on the control of shoot:root ratio, in relation to models. *Ann. Bot.* **61**, 433–449.

Wong, S. C., Kreidemann, P. E., and Farquhar, G. D. (1992). $CO_2 \times$ nitrogen interaction on seedling growth of four species of eucalypt. *Aust. J. Bot.* **40**, 457–472.

Wood, C. J. (1995). Understanding wind forces on trees. *In* "Wind and Trees" (M. P. Coutts and J. Grace, ed.), pp. 133–164. Cambridge Univ. Press, Cambridge.

Zak, D. R., Pregitzer, K. S., Curtis, P. S., Teeri, J. A., Fogel, R., and Randlett, D. L. (1993). Elevated atmospheric CO_2 and feedback between carbon and nitrogen cycles. *Plant Soil* **151**, 105–117.

Zimmermann, M. H., and Milburn, J. A. (1975). Transport in plants. 1. Phloem transport. *In* "Encyclopedia of Plant Physiology," Vol. 1. Springer-Verlag, Berlin.

Index

Physiological Ecology
A Series of Monographs, Texts, and Treatises

Series Editor
Harold A. Mooney
Stanford University, Stanford, California

Editorial Board
Fakhri A. Bazzaz   F. Stuart Chapin   James R. Ehleringer
Robert W. Pearcy   Martyn M. Caldwell   E.-D. Schulze

T. T. KOZLOWSKI. Growth and Development of Trees, Volumes I and II, 1971

D. HILLEL. Soil and Water: Physical Principles and Processes, 1971

V. B. YOUNGER and C. M. McKELL (Eds.). The Biology and Utilization of Grasses, 1972

J. B. MUDD and T. T. KOZLOWSKI (Eds.). Responses of Plants to Air Pollution, 1975

R. DAUBENMIRE. Plant Geography, 1978

J. LEVITT. Responses of Plants to Environmental Stresses, Second Edition
Volume I: Chilling, Freezing, and High Temperature Stresses, 1980
Volume II: Water, Radiation, Salt, and Other Stresses, 1980

J. A. LARSEN (Ed.). The Boreal Ecosystem, 1980

S. A. GAUTHREAUX, JR. (Ed.). Animal Migration, Orientation, and Navigation, 1981

F. J. VERNBERG and W. B. VERNBERG (Eds.). Functional Adaptations of Marine Organisms, 1981

R. D. DURBIN (Ed.). Toxins in Plant Disease, 1981

C. P. LYMAN, J. S. WILLIS, A. MALAN, and L. C. H. WANG. Hibernation and Torpor in Mammals and Birds, 1982

T. T. KOZLOWSKI (Ed.). Flooding and Plant Growth, 1984

E. L. RICE. Allelopathy, Second Edition, 1984

M. L. CODY (Ed.). Habitat Selection in Birds, 1985

R. J. HAYNES, K. C. CAMERON, K. M. GOH, and R. R. SHER-LOCK (Eds.). Mineral Nitrogen in the Plant–Soil System, 1986

T. T. KOZLOWSKI, P. J. KRAMER, and S. G. PALLARDY. The Physiological Ecology of Woody Plants, 1991

H. A. MOONEY, W. E. WINNER, and E. J. PELL (Eds.). Response of Plants to Multiple Stresses, 1991

F. S. CHAPIN III, R. L. JEFFERIES, J. F. REYNOLDS, G. R. SHAVER, and J. SVOBODA (Eds.). Arctic Ecosystems in a Changing Climate: An Ecophysiological Perspective, 1991

T. D. SHARKEY, E. A. HOLLAND, and H. A. MOONEY (Eds.). Trace Gas Emissions by Plants, 1991

U. SEELIGER (Ed.). Coastal Plant Communities of Latin America, 1992

JAMES R. EHLERINGER and CHRISTOPHER B. FIELD (Eds.). Scaling Physiological Processes: Leaf to Globe, 1993

JAMES R. EHLERINGER, ANTHONY E. HALL, and GRAHAM D. FARQUHAR (Eds.). Stable Isotopes and Plant Carbon–Water Relations, 1993

E.-D. SCHULZE (Ed.). Flux Control in Biological Systems, 1993

MARTYN M. CALDWELL and ROBERT W. PEARCY (Eds.). Exploitation of Environmental Heterogeneity by Plants: Ecophysiological Processes Above- and Belowground, 1994

WILLIAM K. SMITH and THOMAS M. HINCKLEY (Eds.). Resource Physiology of Conifers: Acquisition, Allocation, and Utilization, 1995

WILLIAM K. SMITH and THOMAS M. HINCKLEY (Eds.). Ecophysiology of Coniferous Forests, 1995

MARGARET D. LOWMAN and NALINI M. NADKARNI (Eds.). Forest Canopies, 1995

BARBARA L. GARTNER (Ed.). Plant Stems: Physiology and Functional Morphology, 1995

GEORGE W. KOCH and HAROLD A. MOONEY (Eds.). Carbon Dioxide and Terrestrial Ecosystems, 1996

CHRISTIAN KÖRNER and FAKHRI A. BAZZAZ (Eds.). Carbon Dioxide, Populations, and Communities, 1996

THEODORE T. KOZLOWSKI and STEPHEN G. PALLARDY. Growth Control in Woody Plants, 1997

J. J. LANDSBERG and S. T. GOWER. Applications of Physiological Ecology to Forest Management, 1997

FAKHRI A. BAZZAZ and JOHN GRACE (Eds.). Plant Resource Allocation, 1997